Wild Minds

Wild Minds

What Animals Really Think

MARC D. HAUSER

Illustrations by Ted Dewan

An Owl Book

Henry Holt and Company   New York

Henry Holt and Company, LLC
Publishers since 1866
115 West 18th Street
New York, New York 10011

Henry Holt® is a registered trademark
of Henry Holt and Company, LLC.

Library of Congress Cataloging-in-Publication Data
Hauser, Marc D.
Wild minds : what animals really think /
by Marc D. Hauser.
p.   cm.
Includes bibliographical references (p.   ).
ISBN 0-8050-5670-X
1. Cognition in animals. 2. Animal psychology. 3. Social
behavior in animals.   I. Title.
QL785.H359   2000   99-36204
591.5'13—dc21   CIP

Henry Holt books are available for special promotions and
premiums. For details contact: Director, Special Markets.

First published in hardcover in 2000 by Henry Holt and Company

First Owl Books Edition 2001

Designed by Kelly S. Too

Printed in the United States of America

1   3   5   7   9   10   8   6   4   2

For my wife, Lilan, who has enriched my life in countless ways, and taught me a great deal about animal thought as seen through the eyes and hands of a veterinarian

A story must be exceptional enough to justify its telling. We story-tellers are all ancient mariners, and none of us is justified in stopping wedding guests, unless he has something more unusual to relate than the ordinary experiences of every average man and woman.

Thomas Hardy

ACKNOWLEDGMENTS

This book has profited from the insights of science writers, animal lovers, philosophers, veterinarians, ethologists, cognitive scientists, and close friends who were not at all afraid to tell me that I was wrong, and sometimes that I was right and should be courageous. For comments on specific chapters, I thank Kim Beeman, Sally Boysen, Richard Byrne, Susan Carey, Dorothy Cheney, Nicholas Dodman, Gordon Gallup, Gordon Gossage, Don Griffin, Ray Jackendoff, Jerry Kagan, June Kinoshita, Roger Lewin, Peter Marler, Tetsuro Matsuzawa, John Mikhail, Irene Pepperberg, Lew Petrinovich, David Premack, Steven Pinker, Robert Seyfarth, David Sherry, Sara Shettleworth, Elizabeth Spelke, and Andy Whiten. David Premack and Ann Premack also shared several drafts of their own book, and through our correspondences and interactions I have learned a great deal. I dare say that they helped place some corrective lenses on some of my ideas and arguments, and hopefully there is greater clarity now. For reading the entire manuscript with a sharp eye to detail, and for conversational flow, I extend my deepest thanks to Dan Dennett and my wife, Lilan Basse Hauser. Both Dan and Lilan, together with Steve Pinker, have contributed in numerous ways to this book, but especially in making me clear away jargon, clarify arguments, and spice up the stories.

Members of my lab at Harvard commented on chapters during the early phases, and these critiques were extremely valuable in shaping the conversational style of the book. My current students, especially Tecumseh Fitch,

Asif Ghazanfar, Jay Kralik, Cory Miller, Laurie Santos, Dan Weiss, and Mike Wilson, have been an endless source of challenge and fun. Thank you.

My agents Katinka Matson and John Brockman were critical in anchoring my contract with Henry Holt, and have been invested in the project since the start. They have also provided exceptionally fun times at their house in the countryside, a wonderful haven.

David Sobel, my editor at Holt, provided excellent editorial comments on the final draft, pushing me to write a better story and to delete accounts that were less than gripping.

Ted Dewan has brought this book to life with his illustrations. I feel lucky to have his extraordinary imagination expressed through the skills of his pen and ink drawings.

Finally, thanks to my parents, Bert and Jacques Hauser, for allowing me to inherit a sense of wonder, and to my daughter, Alexandra, for being the best kid on the planet.

CONTENTS

All animals are equipped with a mental toolkit. Some tools are universal, shared across all animals, while others are unique.

PROLOGUE

Mental Toolkits

In the summer of 1980 I had a close encounter. I was living in Dade County, Florida. I had come to a tourist spot called Monkey Jungle to conduct research on primate behavior. I also needed to make money, and so tacked on a side job feeding the animals and cleaning their cages.

About halfway through the summer I noticed that a female spider monkey in one cage appeared to be looking intensely at me. I approached the cage. The female spider monkey approached as well. While sitting in front of me, she cocked her head to one side, then the other, and then reached through the cage and slowly wrapped both of her arms around my neck. She looked into my eyes and cooed several times.

What was she thinking?

What was she feeling?

What happened next was remarkable. The female spider monkey's cage mate sauntered over. She looked back, unfolded her hands from my neck, and swatted the male on the head. He jumped back, climbing to a branch on the opposite side of the cage. She then put her arms back through the cage and resumed the position.

I was puzzled. Was this female spider monkey attracted to humans? Or even more worrying, attracted to me? If such feelings were real, was she actually aware of them? And even if she wasn't aware of them, what would this mean for her relationship with her mate? He must surely be infuriated, perhaps even jealous. She eventually let go of her embrace and I walked away.

I approached the spider monkey's cage several times during my stay. She never approached me in the same way. It was a onetime affair.

I suspect that this kind of story may be familiar to many of you. You have, at one point in your life, experienced something like this with your own pet or while visiting a zoo. You have had an animal experience, an emotional encounter with another species.° Furthermore, you probably interpreted the event as if you were inside the animal, assuming that the animal thought and felt as you did. This tendency is natural, compelling, and almost irresistible. It is difficult for us to do anything but interpret animal behavior as we interpret our own.

My encounter with the spider monkey is just one example of a profoundly mysterious phenomenon, one that we experience in our daily lives with humans and with other animals. It is the experience of working out what it's like to be someone else. We seem to have good intuitions about our own species, but can such intuitions be transferred to other animals? Do our pets really feel anxious, happy, guilty, and sad the way we do, or are their facial and bodily expressions simply good copies of what we do without the underlying feeling? When animals find their way home in the dark, are they simply following a trail of odors, or do they refer to a map digitized inside their brain? When animals communicate, are their minds filled with symbols, or are their grunts, coos, and screams the uncontrollable eruptions of passion? Do animals simply follow rules, or do they know why rules are created, why cheaters are punished, and why some actions are right and some are wrong?

This book answers these questions. But the arguments I will develop are different from those that have been offered to date. Many popular books have been written about psychic dogs and cats, weeping elephants, mischievous monkeys, altruistic dolphins, and moralistic apes. Although casual impressions and mythologies about animals are fascinating, they do not bring us any closer to an understanding of what animals truly think and feel. Such wildly speculative accounts, which often leave us feeling good about our warm and furry pets, are counterbalanced by scientists who think that animals are mindless and irrational, driven by instinct, and overwhelmed by their passions. This perspective misses as well because it neglects the now

---

° Throughout this book I use the stylistically cleaner term "animal" to refer to all nonhuman animals, recognizing, of course, that humans are animals too.

substantial research on animal cognition, incorrectly concludes that there can be no thought without language, and fails to place the design of animal minds in the context of relevant ecological and social problems. Animal minds are wild minds, shaped by a history of environmental pressures.

I will show how insights from evolutionary theory and cognitive science have begun to revolutionize our understanding of animal minds. Animals do have thoughts and emotions. To understand *what* animals think and feel, however, we must look at the environments in which they evolved. All animals are equipped with a set of mental tools for solving ecological and social problems. Some of the tools for thinking are universal, shared by insects, fish, reptiles, birds, and mammals, including humans. The universal toolkit provides animals with a basic capacity to recognize objects, count, and navigate. Divergence from the universal toolkit occurs when species confront unique ecological or social problems. Thus, for example, bats echolocate using a high-frequency biosonar signal, but we don't. Unlike humans, bats confront the problem of navigation in the dark. As a result, they evolved a brain that is specially designed to process high-frequency sounds. Humans recognize hundreds of people by their faces, but social insects such as honeybees can't recognize nest mates by their faces. For humans, the face is a special object because it has a unique configuration of features, and because it represents a crucial window into a person's identity, beliefs, and feelings. Consequently, humans evolved a brain that is specially designed to process faces.

The argument I have just developed, albeit briefly, represents the main thesis of this book. The only way to understand how and what animals think is to evaluate their behavior in light of both universal and specialized toolkits, mechanisms of the mind designed to solve problems. And the only way to evaluate the validity of this approach is to test our intuitions about animal minds with systematic observations and well-controlled experiments. Sometimes our tests are carried out in nature, in the environment where mental adaptations evolve. Sometimes, however, the laboratory provides a better environment for testing our intuitions. I will use both approaches in this book.

To illustrate the power of this perspective, I'll return to my interaction with the female spider monkey. Your first impression, like mine, might be that the spider monkey felt affectionate and expressed her feelings by wrapping her arms around me. Set this impression aside and consider a few alternative possibilities. Here is what you would have seen. First, she approached me, put her arms around my neck, and cooed. Second, when the male approached, she removed her arms from my neck, struck him, and then put her arms back around my neck. What, if anything, does this tell us about her feelings? Was she attracted to me in the same way that she would be attracted to another spider monkey? Let us start by looking at how female spider monkeys act toward male spider monkeys when they are attracted to them. Do they sidle up, put their arms around their mate, and coo? If so, what physiological changes do they experience? When we are romantically moved, our insides are excited as heart rate increases and hormones rage. Many animals may experience similar physiological changes, and we can certainly measure them. We can't, however, assume that the underlying physiological changes lead directly to the highly subjective feelings that humans associate with being "in love" the way Romeo and Juliet were, or Tristan and Isolde, or Ryan O'Neal and Ali McGraw. But we can begin to test this assumption by comparing the female's physiology during her encounters with me and during interactions with a male of her kind.

Assume that the monkey's physiology is the same when she is with me and when she is attracted to a male spider monkey. We have learned something important, but we don't know whether she is attracted to me as a mate, as another animal who is nice to her, or because I am the best keeper she has ever seen. To distinguish between these alternative interpretations, we need to make other observations and tests. For example, we can start by assuming that cleaning her cage is less important than feeding her. Would she be as friendly toward a keeper who cleaned her cage but never fed her? If so, then her response is not dependent on the feeding.

What can we make of her aggressive gesture toward the male? Was it related to her attraction to me? Under what circumstances does she attack her mate? Does she strike him whenever she is engaged with someone or something else? For example, if she is feeding and he approaches, does she strike him? If so, then her aggressive attack may have more to do with dominance than with her relationship to me. What if a new male spider monkey enters the cage? Would she be more interested in him than me? Am I just the new boy on the block? Are her acts guided by a desire for revenge or to enlist jealousy? What about the possibility that a previous keeper trained her to respond in this way? In the past, whenever she approached, the old keeper gave her food. One day she accidentally put her hand on his shoulder and he reinforced her with a special treat in much the same way that animal learning psychologists such as B. F. Skinner used food reinforcement to train pigeons to classify stimuli. Over time, the keeper shaped the spider monkey to put both hands around his neck. In the end we have what appears to be an affectionate spider monkey. In reality, however, what we have is a monkey whose every action has been shaped by the keeper—a puppet on a string. Although there may be some feeling, it is unlikely to be of the kind that we originally guessed from her response.

This story forces us to appreciate that for any given behavior there are many possible causes, and thus many possible explanations for why a particular action is taken. Like our pets, the spider monkey may have felt a moment of affection toward another species, but maybe not. The problem we face is to figure out what kinds of feelings and thoughts animals have, and why they evolved such capacities.

In contrast to most books on animal thought and emotion, the ideas I develop here depend critically on recent findings in the neurosciences and studies of human infant development. Studies of the brain, which can be explained without technical jargon, are critical for our exploration of the animal mind and its evolution. By understanding some functional anatomy—what different parts of the brain do—we learn a great deal about the limits on animal thought. Several authors claim that animal thought is limited or nonexistent because animals lack language. This claim assumes that language is necessary for thought, and that animals and *adult* humans are the most interesting groups to compare. In light of recent findings in the study of animal cognition, I argue that language is not necessary for certain kinds of thought, and that the most profitable comparison among species is between animals and human *infants*. Although human infants are born with a brain that is designed to process language, it takes years for them to develop an adult level of competence. Consequently, the most profitable approach is to compare the mental tools of human infants and animals, and then explore how the minds of infants are transformed by the acquisition of language, a new tool.

In the first part of the book you will encounter some challenges that all animals confront, such as recognizing food as a particular kind of object, assessing the number of individuals in a neighboring group, and finding the shortest path home. These challenges are solved by the universal toolkit, specialized mechanisms for recognizing objects, counting, and navigating. In the second part of the book we'll get to the heart of our discussion, the mental tools required for psychological knowledge. Here we explore whether animals have a sense of self and the capacity to imitate and deceive each other. Only a few species in the animal kingdom have added these mental tools to their toolkits. In the final part of the book we explore how animal societies are organized, changed, and kept in balance by rules, regulations, conventions, and styles of communication. The work reviewed shows how animals use their mental toolkits to make a living in their social and ecological worlds, and how humans, alone, have applied their tools to the creation of a moral world.

The following series of questions and answers will inform our discussion.

• Do animals *think*? Are animals *conscious*? Are some animals more *intelligent* than others?

I think these are unhelpful questions because they are vague, relying on general concepts that are often defined on the basis of what humans do. In this spirit, I will generally avoid using the words "think," "conscious," and "intelligent." Instead, I will ask about mental phenomena that are more precisely specified, phenomena such as an animal's capacity to use tools, to solve problems using symbols, to find its way home, to understand its own beliefs and those that others hold, and to learn by imitation. These are issues that I believe can be tackled independently of human thought and with scientific methods.

• Do animals have emotions?

Yes. Emotions prepare all organisms for action, for approaching good things and avoiding bad things. But when we step away from the core emotions such as anger and fear that all animals are likely to share, we find other emotions such as guilt, embarrassment, and shame that depend critically on a sense of self and others. I will argue that these emotions are perhaps uniquely human, and provide us with a moral sense that no animal is likely to attain.

• Do animals communicate?

Yes. But each species' communication system has unique design features, specialized for transmitting information and manipulating behavior. Thus, bats and dolphins echolocate, birds sing, honeybees dance, gorillas grunt, kangaroo rats drum, and fireflies flash. And yes, humans speak and gesture in their native language, and sometimes in a few foreign languages as well.

· Are animals guided by instinct?

Yes, and so are we. Instincts guide the learning experience. Instincts cause organisms to attend to some features of the environment and ignore other features. Many of our instincts are evolutionarily ancient, causing us to share a perceptual and conceptual view of the world with many other animals.

· Do animals have rules by which they abide, and sometimes break?

Yes, and the rules reflect the conditions under which the games of reproduction and survival are played out. But unlike our rules, theirs are not based on some understanding of what is "right" and "wrong." Animals obey rules, but don't know that rules have been designed to preserve conventions, to prevent harmful actions, and in some cases at least, to maintain fairness. Human infants are just like animals, but human adults are not. As adults, we know that rules were designed to mediate between right and wrong, and that rules provide guidelines for action. Humans consult rules, and often obey them. More important, because of our capacity to distinguish between "is" and "ought," between the biological biases that we have inherited from our past and what we decide to do as individuals while constructing a life, we are uniquely positioned to set values on behavior, to reward certain actions and punish others.

In addressing these questions, I draw on my experience as a scientist studying animals, in the wild and in the lab, but also as a casual naturalist. I have observed and run experiments on animals including vervet monkeys living on the savanna of Kenya, chimpanzees in a Ugandan rain forest, crows on a golf course in southern California, white-crowned sparrows overlooking San Francisco, rhesus monkeys on a tropical island off the coast of Puerto Rico, and cotton-top tamarins in my lab at Harvard University. This provides me with countless observations of animal encounters that I frequently weave into the conceptual texture of the issues at hand.

Understanding an animal's view of the world requires, therefore, an ability to isolate the appropriate sensory systems and determine the kinds of problems that the species surmounted in the past in order to survive into the present. We must not assume that we know what they know, or how they have come to know it. Unlike Scrooge, who was given the opportunity to visit his past and look into his future, we must use the current design of an animal's brain and behavior to infer the kinds of problems that its mind was designed to solve. This is a recipe that has worked since Charles Darwin cooked it up over a hundred years ago. If we follow it closely, we will learn a great deal about the wild minds that inhabit this planet.

Wild Minds

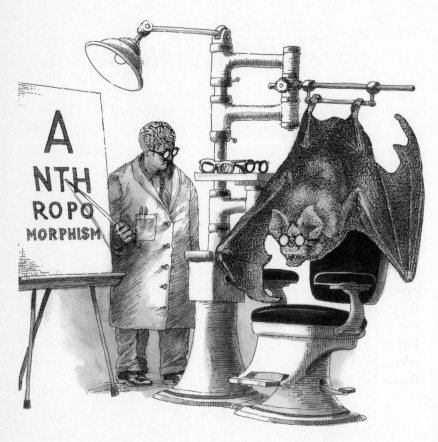

Skeptical scientists, blind bats, and the dangers of anthropomorphism

1

Animal Tales

In one of Rudyard Kipling's charming short stories, he tells us about an elephant child with insatiable curiosity, an elephant who asked so many questions that his relatives spanked him. Curious minds, like the elephant child's, abhor ambiguities, feelings of ambivalence, and the lack of resolution. But curious minds discover uninhabited terrains of knowledge by questioning dogma and pondering the impossible. In the novelist Edith Wharton's words, if we remain "insatiable in intellectual curiosity, interested in big things, and happy in small ways," we will remain alive. Curiosity exacts a cost, but the returns are great. Although animals may not be as curious as Kipling's elephant child, they are active informavores, digesting and storing relevant information in the service of guiding behavior.[1] Let me illustrate this idea with a few vignettes.

   • When a common laboratory rat is placed in a maze, it immediately begins to explore, both with its nose and with its eyes. Waiting for the lunchtime food truck to arrive or for the exit sign to be illuminated won't help. With exploration comes detailed knowledge of the turf, an understanding of which way to go for food and which way to go for the exit. Curiosity allows the rat to create a road map, a directory of spatial coordinates.
   • Chimpanzees housed in outdoor enclosures often discover novel escape strategies. They insert sticks into the fencing for

footholds. Some are even more creative, placing large logs at an angle against the enclosure wall, creating functional ladders. By assessing the height of the enclosure wall and trying out different logs, the chimpanzees eventually escape, thereby discovering whether the grass is, in fact, greener on the other side.

· In many species of schooling fish, individuals leave the safety of their group to swim by and inspect the behavior of a nearby predator. Rather than wait for the predator's attack, such bold inspectors gain information that can be used to decide whether to stay or flee. In fact, they not only gain information, but relative to those who tag behind, lower their odds of being eaten.

· Whenever you take a dog out on a walk, even on a route that it has taken on every outing, the dog sniffs the ground, trees, and fire hydrants. Dogs are interested in the competition. Sniffing allows them to extract scents that other animals have left behind. And of course, they always sign off with their own unmistakable signature.

The claim that animals are curious is based on the general assumption that by looking at the similarity between animal and human behavior we can make similar inferences about their thoughts and emotions. Although our intuitions may sometimes be correct, behavior *can* be a misleading guide. It is appropriate to be a healthy skeptic, and this chapter shows why.

ANIMALS R US

A number of popular writers theorize that animals and humans think about the world in the same way. These authors also portray scientists as the enemy, naysayers and skeptics. For example, in her preface to *The Hidden Life of Dogs*, the essayist Elizabeth Marshall Thomas begins,

This is a book about dog consciousness. To some people, the subject might seem anthropomorphic simply by definition, since in the past even scientists have been led to believe that only human beings have thoughts or emotions. Of course, nothing could be further from the truth. . . . [W]hile the question of animal consciousness is a perfectly valid field for scientific exploration, the general assumption that creatures lack consciousness is astonishing. . . . After all, thoughts and emotions have evolutionary value. If they didn't, we wouldn't have them.

Thomas makes two errors here, the first concerning evolutionary continuity, and the second concerning adaptation. Given that humans diverged from their chimpanzee-like ancestors five to six million years ago, it is possible for us to have evolved capacities that other animals lack. No miracles. We cry with tears, blush when we are embarrassed, and walk bipedally. Our primate cousins don't. And although consciousness may be adaptive in humans, we cannot deduce from this conclusion that animals must also be conscious. Further, it is a mistake to argue that anything with apparent "evolutionary value" must be an adaptation and therefore should have evolved in animals. A lion out on the savanna might do well with wheels instead of legs, but such a system cannot evolve due to constraints imposed by the nervous system. Wheels would cause a lion's neurons to wrap up into a tangle of spaghetti.

Jeffrey Masson, another popular writer about animals, is a Sanskrit scholar who is best known for his attack on Freud and the psychoanalytic tradition. He opens his book *Dogs Never Lie about Love* with a similar comment about scientists, science, and the study of animal behavior:

> I am aware that most of the "evidence" I have presented for the reality of emotions of dogs consists of stories—what scientists call, dismissively, anecdotal evidence. With their restricted sense of valid criteria, most scientists want to be able to test, probe, and replicate data. You cannot do that with a single story. Scientists seem to think that whereas a story can be either true or false, something that takes place more than once in a laboratory has to be true. There is no reason to believe this to be the case. Data can be faked, forged, or misrepresented as easily as can a story, and what we learn from some laboratory experiments . . . does not tell us anything we could not have known without experiments.

Masson's criticisms are replete with disturbing and confusing assertions. Scientists think that anecdotes are unsatisfying, but not useless. They may provide clues or stimulate hunches, but scientific curiosity ultimately leads to further exploration, additional observations, and experimental tests. And yes, scientists can fake data, which is precisely why replication is so important. For example, in 1989 a team of physicists claimed they had evidence of cold fusion, a process in which the fusion of two heavy hydrogen nuclei generates energy. Subsequent attempts to replicate these findings failed, showing instead that the energy generated from the original experiments was due to stored heat within the general system, rather than heat released from the fusion of two nuclei. Had we invested in a cold fusion mutual fund, we would now be deep in debt. In the remaining chapters of this book, I will show how experiments have enabled scientists to avoid incorrect conclusions and gain revolutionary insights into the animal mind. But first, let me try to convince you that animal stories are unsatisfying.

Chimpanzees Form a Natural Bridge

In 1987 I was in the Kibale Forest of Uganda watching a small family of chimpanzees, a mother, her subadult son, and her one-year-old daughter. After feeding in a fig tree for a while, the subadult male gave a departure call and leaped across a gap in the canopy to a tree some distance away. The mother soon followed, but her daughter stayed behind, screaming. The mother and son waited and watched, but the yearling appeared frozen in place. After a stalemate, the mother went back to her daughter, started swinging the tree back and forth, and then, with a long stretch, reached over and grabbed the branch of the neighboring tree. With her feet grasping one tree and hands grabbing the other, the mother formed a natural bridge, which her daughter used to cross over to safety.

What I witnessed was magical and immediately invoked a suite of questions concerning maternal care. How often do chimpanzees create natural bridges? Do they create a mental image of their body bridging a gap in the trees before actually stretching across the canopy? Do they create bridges for any yearling, juvenile, or adult in need? How does an individual recognize another in need? Does a mother empathize with her daughter when she is stuck behind, screaming? Would she empathize with an unrelated yearling frozen in the same position? To address these questions, we would need to make additional observations. The insistence on replication is not a silly scientific ritual, performed by priests in white lab coats. It is a tool for understanding whether an event is common or rare, and why it occurred. In this case, the mother's actions appear to be intentional and deliberate. But did she actually plan to make a bridge because she knew it would allow her daughter to cross the canopy? If so, did she invent this technique or did she learn it from her mother, who learned from her mother, a tradition passed down through the ages? Perhaps she was merely showing her daughter how to cross. In the midst of the mother's demonstration, however, the yearling seized the opportunity and walked across her mother's back. If the second description is correct, then we must seek a different interpretation. Either the infant hit on an insightful stroke of genius or she was blessed with dumb luck.

Returning to Masson's statement, it is not the case that such "stories" are uninteresting. Indeed, they are fascinating, but limited with respect to what we can learn from them. One observation leads to several intriguing questions and problems. The only way to address them is by collecting additional observations and, if possible, running experiments. Of course, that is not always possible. We can't, for instance, necessarily induce our bridge-forming mother to do so again. We may wait for a long time for another occurrence. Even if we can induce others to form bridges, what would that tell us about the emotions of the bridge makers and the bridge users? How can we understand whether each bridge maker invented the technique on her own or learned the technique by imitating someone else? A second example might help answer some of these questions.

Compassionate Dogs

In *The Hidden Life of Dogs* Elizabeth Marshall Thomas uses observations of dog behavior to make claims about their emotions and thoughts. Her interpretations, however, are overinterpreted and fraught with assumptions. Here is an example:

> [M]ost animals, including dogs, constantly evaluate other species by means of empathetic observation. A dog of mine once assessed my mood, which was dark, over a distance of about one hundred yards, and changed his demeanor from cheery to bleak in response.

To tap into your own intuitions about Thomas's observation and interpretation, try the following thought experiment. Substitute another animal for the word "dog" in the passage above and reread it. For example, "A *lizard* of mine once assessed my mood . . ." Play with this thought experiment until you hit on an animal that, when inserted into the sentence, makes it sound absurd. Even if you never come upon such an animal, and my hunch is that you will, what assumptions are you making about these animals' thoughts and feelings? Let's say that you are convinced that a dog can read someone else's mood, while a slug, lizard, or fish cannot. You are making a distinction somewhere in the tree of life, a distinction based on what you can observe. As far as you can tell, zucchinis lack thoughts and emotions. Whatever thoughts and emotions you ascribe to slugs, lizards, and fish, you may not believe that they are capable of creating a bond with their owner. Fine.

Let's return to Thomas's description and analyze what she observed. The dog spots Thomas at a distance and is bouncy, tail wagging. As he approaches and looks at Thomas, his head and tail drop and he loses the bounce in his walk. How are we to interpret what the dog is feeling? Thomas's interpretation assumes that dogs empathize with humans. To empathize, however, the dog must not only understand what human feels, but feel the same as well. Thomas's dog must experience Thomas's dark mood. Alternatively, perhaps the dog is simply responding to Thomas's behavior without feeling what she feels. He has lost his bounce because he correctly anticipates, based on prior experience, a less than affectionate response from Thomas. When she is smiling, the greeting involves play and lots of back scratching. When she is frowning, the greeting is brief and dismissive. The dog's dark mood arises, therefore, out of selfishness rather than empathy. Thomas's dog has lost an opportunity for quality time.

Thomas's example and the interpretation she offers are a classic case of seduction. We are seduced by appearance. If there are familiar cues or signals, we tend to ascribe similar emotional experiences and thoughts. But appearances can be misleading, guiding us down mirrored halls.

Since 1988 I have been studying rhesus monkeys on the island of Cayo Santiago, located just off the coast of Puerto Rico. During the breeding season, males fight a great deal. Dominant males often chase subordinates away from potential mating opportunities. Sometimes the dominant catches a subordinate, pins him down, and rips out one of his testicles. While the testicle is being ripped out, the subordinate is generally quiet and there are no noticeable facial expressions. Even more striking, the injured male is often seen mating or attempting to mate within a few hours of being attacked. From what we observe, how can we determine what the subordinate is feeling? Can we really use our own experiences, imagined or real, to deduce what subordinate rhesus monkeys feel in this situation? Unlikely. Although appearances often provide us with accurate tips for what is going on inside another individual, we must tread cautiously.

Jeffrey Masson describes a case of "compassion" in dogs. It is, on the surface, similar to Thomas's example of empathy and my own example of the chimpanzee bridge.

> When one of my three dogs strays too far from the others and I continue walking, oblivious, I will notice that the other two stop and wait for their companion to return. . . . They do not want to continue until the pack is complete. This act is surely indicative of compassion. . . . Of course, we could explain it in other ways; there is always another explanation, whether for human or animal compassion. . . . But even if there is some truth in these explanations, they do not cancel out the element that derives from love and compassion because they cannot explain away the feelings that accompany those actions.

The act of waiting is interpreted as a sign of compassion. From this interpretation, Masson makes the point that even if other factors such as self-interest can account for the act, this doesn't negate the importance of emotion in guiding behavior. What Masson states here is important at one level, and wrong at another. It is important because there are always different ways to explain why something occurred. It is wrong because although all behaviors can be explained in different ways, one kind of explanation need not exclude the others. The explanations are not mutually exclusive.

Why do dogs move in packs? From the perspective of understanding the *mechanisms* linking cause and effect, we might say that dogs form groups and stay together because they have a desire to affiliate, a desire that is presumably driven by the anticipation of feeling happy or something like it. In many animals there are parts of the brain involved in the regulation of attachment between individuals. Highly social animals, such as dogs, monkeys, and humans, undergo hormonal changes associated with separation that are different from those experienced by less social animals. From a *developmental* perspective, we can explain grouping as something that is instinctual, a genetic predisposition that is shaped by the experience a puppy obtains from watching siblings and its mother. At an *adaptive* level, concentrating on traits that promote increased survival and reproduction, we might argue that natural selection favors animals who stay with the pack because those who stray are more likely to be attacked and killed by predators. Although waiting may be costly, it may represent a kind of altruism that is based on reciprocation—I wait for you today, knowing that you will wait for me tomorrow. Using a *phylogenetic* or *historical* analysis, we would conclude that staying in packs represents an evolutionarily ancient characteristic, one that is seen in the domestic dog's ancestors, the wolves. Pack formation is a characteristic that domestic dogs inherited from wolves.

We gain a rich understanding of behavior by addressing each of these four causal questions. This approach was developed in the 1950s by one of the founding fathers of the study of animal behavior, the Nobel laureate Nikolaas Tinbergen.[2] To this day, it represents the foundation of modern research in animal behavior.

What's all the fuss about then? Few students of animal behavior would argue that a dog, or any other animal, lacks emotion. Disagreement enters the discussion when we attach a certain level of confidence to the kind of emotion based on anecdotal observations alone. Waiting for a pack member to catch up could be a sign of compassion or it could be a sign of fear, anxiety, the desire to control pack movement, to find out what the stray learned on his brief excursion, and so on. To discover what a dog or any other animal feels and thinks, we must carry out systematic observations and experiments, guided by the theories of evolutionary biology and cognitive science.

PETS ON THE COUCH

Nicholas Dodman, a veterinarian at Tufts University, has spent his career working on pet dogs and cats with behavior problems. He opens his book *The Dog Who Loved Too Much* with the claim that people may "have vague images of a voodoo-style veterinary psychiatrist who will put their dog on a couch and inquire about its puppyhood. Some owners even fear being psychoanalyzed themselves." But this is no pet quackery. Dodman discusses a growing body of work indicating that many of the behavioral problems that dogs experience can be treated with the same type of pharmacological drugs that work on human psychiatric disorders. In particular, some dogs manifest behavioral symptoms indicative of bipolar disorder, depression, separation anxiety, dominance aggression, rage, and obsessive-compulsive disorder. In some cases, the problems appear to be breed-specific: anxiety-prone Afghans, springer spaniels with rage syndrome.[3] Consider the following case of obsessive-compulsive disorder in a Doberman named Taylor.

> Taylor . . . like many Dobermans, had a licking problem. Taylor was first noticed to be somewhat orally inclined when he was about one month of age; he would knead cushions and blankets with his paws, mouthing and sucking the items until they were wet. . . . As a puppy, [he] would also nurse on other dogs' ears, and when there were no blankets or ears around, he would just suck and chew stones or mouth the skin of his flank until it was moist with saliva. Blanket sucking and flank sucking occur almost exclusively in Dobermans, suggesting that the condition may be genetic. . . . When Taylor was two, he started another oral activity: licking his legs.

There are several intriguing observations and intuitions here. The repetitive licking of one spot in early development, and the migration to other body parts later on, is reminiscent of the many obsessive-compulsive cases reported for humans. In particular, the disorder has a strong genetic component, often expressed at a young age, and most individuals tend to manifest similar behavioral symptoms, usually excessive washing. Later in life, the problem becomes highly stylized and personal, with some patients driven to repeatedly checking that the lights in a room are off, or walking in and out of a door frame until they cross at the precise midpoint. In one famous case, a young man with an extreme hand-washing problem was driven to an attempted suicide. He took a gun and shot himself in the mouth. Rather than ending his life, however, he actually fixed the problem. The bullet went straight through the frontal lobe of his brain—the prefrontal cortex—causing a lesion and ending the obsessive-compulsive disorder from which he suffered.[4] As some researchers suggest, patients with obsessive-compulsive disorder clearly experience difficulties when an action must be inhibited. Several studies of humans and animals indicate that the prefrontal cortex plays a critical role in the process of inhibition. By shooting himself, the patient was lucky enough to remove the troubling area of his brain. As they say, this kind of self-help treatment is not recommended.

Are there other, less invasive techniques for treating obsessive-compulsive disorders? Following recent advances in human medicine, Dodman and other veterinarians are pursuing a course of treatment that involves both behavioral therapy and drug administration. For example, repetitive licking in Dobermans often emerges in response to boredom and stress, and with time becomes highly uncontrolled and repetitive. Behavioral treatment involves increasing the level of exercise, altering the diet, and putting a collar around the neck to prevent licking. Much greater success is achieved when behavior therapy is combined with drugs that directly affect the neurotransmitters, the chemical messengers of the brain. Although there are significant differences in brain anatomy between animals, there are relatively few differences in their neurotransmitters.[5] Consequently, neuropsychological disorders in humans and dogs can be treated with many of the same drugs.

Dodman's work provides one approach to the problem of comparing animal psychologies. Under certain conditions, dogs and humans appear to show similar behavioral responses and manifest the same sorts of neurochemical changes. The work is elegant and sits properly within the tradition of Sherlock Holmes and the scientific method. One observation provides a clue. Additional observations are then collected so that hypotheses can be formed and tested through experimentation, in this case, therapy, drugs, and further observation. Unfortunately, a problem remains. Although the behavior and neurochemistry are similar, this doesn't guarantee that the intervening thoughts or feelings are the same.

Consider the following hypothetical example involving two human patients diagnosed with obsessive-compulsive disorder. Whenever they are presented with an array of objects, they each sort the square ones from the others, then put the objects back in a pile and start over again. At the end of the sorting process, both patients have a pile of square objects and a pile of nonsquare objects. Their behaviors are the same and previous tests reveal that the underlying neurochemical imbalances are the same. The problem is that when we ask them about their sorting rules, one states that he was looking for parallelograms, whereas the other says he was looking for squares. By chance, it turns out that all the parallelograms around are squares. As a result, although their piles look identical, each patient used a different conceptual representation while sorting the squares from the nonsquares. Same behavior, same neurochemistry, but different thoughts.

ANIMAL PARAGONS

Serious shoppers want the best deal. They research the field, identify the competitors, list all the pros and cons, and then make a decision based on their priorities. It's a game of compare and contrast. Humans, and many other species, face a similar problem in finding the best mate. In the story of Cyrano de Bergerac, poor Roxanne thinks that she is in love with the handsome Christian. When she finds out that the poetic spirit behind his voice is, in fact, the physically unattractive Cyrano, she is forced to make a choice between physical and poetic beauty. No easy answers here, but there is a choice. Shoppers for mates, just like shoppers for clothes, microwave ovens, or cars, must be able to dig beneath the superficial gloss. There is rarely truth in advertising.

The history of research on animal thinking has, in many ways, been driven by a shopper's mentality: to find out who is the real brain, the whiz of the animal kingdom. And the comparison is, most commonly, between humans and the other animals. As the philosopher Hilary Putnam stated, "Our world is a human world, and what is conscious and not conscious, what has sensations and what doesn't, what is qualitatively similar to what and what is dissimilar, are all dependent ultimately on our human judgments of likeness and difference."[6] Poets, novelists, and scientists have identified us as superior because we are the noble, rational, intelligent, social, successful, creative, tool-making, linguistic, and anxiety-driven species. We are also, as Mark Twain noted, "the only animal that blushes. Or needs to."[7] No one expresses this overall vision of superiority and uniqueness more eloquently than Shakespeare's Hamlet, who called humans "the beauty of the world, the paragon of animals." To admire our species for its qualities is natural. To place us with the gods and angels, *above* all the others, is both pompous and boring. It is pompous because it places us on top of an intellectual pyramid without articulating the criteria for evaluation. It is boring because it ignores differences in thinking, and fails to search for an understanding of how different shades of mind evolved.

A commonly expressed dichotomy is that animals are driven by their passions whereas humans are guided by reason. We are rational, coolheaded, and thoughtful. They are irrational, hotheaded, instinctual beasts. Whereas we follow the dictum "I think, therefore I am," they adhere to "I feel, therefore I act." The division between the emotions and rationality represents a vestige of Cartesian thinking. As the neuroscientist Antonio Damasio pointed out in his 1994 book *Descartes' Error*, our emotions often play a critical role in guiding our decisions. When we attack a problem, we often consult our emotions, using them as guides. Although it is clear that our emotions can cloud and overpower our rational minds, we are more often than not making decisions that depend on an understanding of what we feel. In John Le Carre's novel *The Taylor of Panama*, Marta, an upstart in the student revolution, states the case with precision: "Reason only functions when the emotions are involved. . . . There's no logic unless the emotions are involved. You want to do something, so you do it. That's logical. You want to do something, and don't do it, that's a breakdown of reason."[8]

The idea that rationality depends on the emotions comes from studies of brain-damaged patients and the use of modern neuroimaging techniques such as petscan (PET) and functional magnetic resonance imaging (fMRI). The capacity to make rational decisions breaks down when the planning area of the brain (prefrontal cortex) is disconnected from a key emotional area of the brain (amygdala). Further, neuroimaging results show that these two areas of the brain are highly active when non–brain-damaged humans make decisions about emotionally salient problems. The conclusion we must then draw from this work is that emotions are necessary for making decisions in humans. What I will argue is that emotions play a central role in animal decisions as well. Many animals, especially the nonhuman primates, have a prefrontal cortex, an amygdala, and the neural circuitry that connects these areas. This neural gear provides them with rich emotions and a foundation for making emotionally relevant decisions. Without delving into the details, I leave this problem for later chapters and simply claim that Descartes' dichotomy is a nonstarter. We need a better theory for understanding animal minds.[9]

Darwin's theory of natural selection—and the logic of design and engineering that it encompasses—is central to the thesis of this book. Just as a variety of transportation vehicles have been engineered, natural selection has engineered brains with specialized devices, what I call *mental tools*. Some tools are shared across species, whereas others are unique. Whether the design of a mental tool is good or bad is decided in a competitive arena, one in which only the winners survive, reproduce, and pass on genes into the next generation.

Consider the fact that once the wheel was invented as a transportation tool, its functionality depended on the invention of roads. Once roads were in place, increasingly sophisticated land vehicles emerged, with horse and buggy outpaced by Henry Ford's Model T, followed by the Lamborghini, and rocket-like hot rods that can now break the sound barrier. Travel by air and water presented different challenges, but technology responded with the invention of sailboats, kayaks, windsurfs, jetskis, gliders, airplanes, and rockets that can take us away from our beloved planet Earth. Some of these designs were elegant and relatively efficient. With the march of time and technology, however, the state of the art changes, each generation contributing design improvements as well as new tools.

Like our man-made vehicles, animals also confront problems of locomotion on the ground, in the air, and through the water, and their brains have been built to solve such problems, tracking changes in topographic contours, temperature gradients, and wind and water currents. But animal brains must cope with other problems as well. Brains have been designed to help animals find a meal, avoid becoming one, pick a mate, care for offspring, build friendships, learn from experience, communicate, remember where home is, avoid enemies, collaborate with allies, and much more. Like transportation vehicles, specializations of the animal brain emerge and persist in the face of unique but recurrent problems. Similarities in design arise when the set of possible solutions to a problem is limited. Let me illustrate these points with an example from nature's favorite drama, the epic battle between predator and prey.

The Firefly's Fatal Flash

Tennyson's song "Now Sleeps the Crimson Petal" opens with a stanza about fireflies whose magical light gently awakens the sleeping plants and people. Little did he realize that the flash of the firefly represents both an invitation to mate and a neon advertisement to predators. When females of the predatory *Photinus* species see the courtship flash of a male from the prey species, *Photuris*, they respond with a deceptively accurate copy of a flash from a female *Photuris*. This visual wink from an interested female is sufficient to lure the male in. When he arrives, she calmly devours him. Lust can sometimes kill a male firefly.

There is a fascinating twist to this firefly story, described in 1997 by the biologist Tom Eisner and his colleagues at Cornell University. As in many complex environments, animals often experience life as both predator and prey. *Photinus*, the predatory firefly, is eaten by thrushes and jumping spiders. *Photuris*, in contrast, is less vulnerable to thrushes and jumping spiders thanks to a plant it eats that contains a chemical the predators find unpalatable. Following predation on *Photuris*, the predatory *Photinus* not only boosts its energetic reserves, but also simultaneously ingests the same chemical that makes *Photuris* unpalatable to jumping spiders and thrushes.

What does this epic battle tell us about the design of animal minds, and in particular, firefly minds? From these observations, we cannot conclude that females of the predatory firefly species are aware of what they are doing. Presumably, they didn't sit down and calculate the effectiveness of the plant's chemistry and its role as a prophylactic against predators. Rather, natural selection favored the capacity of female *Photinus* to recognize the mating flash of *Photuris*, signal a deceptively seductive flash saying, "Yes, I'm interested," and then devour the courting male, thereby obtaining a meal and a vaccine against predation. For this system to have evolved, encounters between firefly prey and their predators—jumping spiders and thrushes—must have been relatively frequent, statistical regularities of their environment.[10]

Does the predatory *Photinus* deserve a genius award for outwitting its prey, *Photuris*, as well as the thrushes and jumping spiders? On an evolutionary time scale, *Photinus*'s success may represent a fleeting victory, an eye blink of glory. Most predator-prey battles operate like military arms races. Predators win for one stretch of time, and then prey evolve defenses to evade predation, thereby gaining the upper hand in the evolutionary race. For example, male *Photuris* capable of recognizing mimics would be favored. Similarly, jumping spiders capable of discriminating *Photinus* females that have recently eaten *Photuris* males from those that have not would be favored by natural selection, although there may well be constraints for evolving such abilities; for example, because spiders can't continuously monitor the food intake of *Photinus* females, they would have to use a secondary cue such as smell to determine what had been eaten.

Though *Photinus* has been equipped with a smart hunting device, it is only smart in a narrow sense. If *Photuris* changed the signature of its flash, it seems unlikely that *Photinus* would be able to mimic in kind, at least not over a short period of time. There is no evidence that the mimetic system is flexible, capable of matching any variant that comes its way. The capacity represents a highly specialized skill, one that is radically different from our own capacity to mimic or imitate (see chapter 6).

Fireflies are not the only mimics in the animal kingdom. But when we compare across animal groups, we find that there are differences in the mimetic tricks themselves as well as in their functions. In some species the trick is fixed, part of the animal's body. In other species it's like a Halloween costume, a façade, one that can be put on and then taken off depending on the season. Thus, some songbirds such as the chaffinch and lyrebird can mimic the sounds of other birds and even nonbiological sounds such as the ring of the telephone. One explanation for mimicry in songbirds is that it provides an illusion that a given habitat is saturated—a vocal "No Vacancy" sign. The bird mimics by listening, storing, and then reproducing the sound heard. Mimicry is a capacity that is based on innate mechanisms for learning about particular sounds in the environment.

Some snakes, when confronted by a potential predator, will mimic the posture of a dead snake, turning their head upside down while keeping their mouth wide open. Since predators typically bypass dead things, the snake's death-feigning performance represents an effective ploy, a functional case of pretense. A similar ploy is used by the burrowing owl. Because burrows are scarce and highly valuable, the owl has evolved a high-fidelity copy of the rattlesnake's rattle in order to fend off potential intruders from its burrows. In contrast with songbird mimicry, which depends on specific experience, the death-feigning performance of the snake and the owl's rattle emerge without practice, experience, or instruction.[11] As we can see from these examples from nature's directory of Marcel Marceau acts, what is similar at one level can be quite different at another.

The following chapters show how evolution has placed specialized mental tools into the toolkits of species as diverse as the Tunisian desert ant, the New Caledonian crow, the Indian rhesus monkey, and the urban human infant. Each chapter, then, represents a piece to an evolutionary puzzle, a puzzle about the construction of animal minds. As Aristotle so elegantly put it, "In all things of nature there is something of the marvelous."

PART I

Universal knowledge

The first installment in the universal toolkit is a mental tool that allows all animals to recogniz[e] objects and predict their behavior. Each animal, however, uses this mental tool in the service o[f] unique social and ecological problems.

2

The Material World

Imagine that you are a primatologist following chimpanzees in the forests of Sierra Leone. Your target subject, an adult female, approaches the base of a kapok tree filled with large ripe fruits. As with many fruiting trees, the kapok is armed with physical defenses in the form of long sharp thorns. But your previous observations indicate that this community of chimpanzees has a thing for kapok fruits. Your subject will almost certainly climb the tree, trying to avoid the thorns. But, as so often happens when we watch animals, your target female surprises you. She rips off a smooth branch from the tree and places one piece under one foot and another under the other foot. With her toes securing the branch to the soles of her feet, she climbs the tree, marching over the thorns like a steamroller. When she reaches a patch with an abundance of fruit, she strips off another smooth branch and places it down over some thorns so that she can sit in peace. Comfortably seated, she begins to feed.

This is a true story, based on observations made in 1997 by the primatologist Rosalind Alp.[1] Not only did she observe one chimpanzee using such sandals and seats, she observed several individuals in the community doing the same. To both invent and use these tools, the chimpanzees must have more general knowledge of physical objects, of the kinds of objects that work as sandals and seats. Thus, for example, they must understand that some objects are suitable for crafting sandals and seats, but others are not. If so, then they should choose a pair of Nike hiking boots with crampons over a branch because the Nikes represent an optimal design for the task at hand. They should also realize that branches may break, but their feet can't go through them, and once attached, won't vanish in thin air. Although these may seem like reasonable claims, we can't be certain. We must test the claims that "they must understand" by exploring how object knowledge is acquired in animals, and the extent to which such knowledge guides their use of tools.

When we create an artifact such as a tool, we leave a physical trace of our thoughts. The creative act involves imagining a design and then planning a series of actions that will lead to its construction. But it is not always easy to discern what the designer intended to create. Sometimes an artifact's form fails to give away its function.

In the movie *Sleeper*, Woody Allen's character is brought back to life after being cryogenically frozen for two hundred years. Upon regaining all his mental faculties, he sits down with a scientist who wants to determine the function of artifacts dated to the 1970s. When shown a set of chattering teeth, Allen says that they were used as party gags to make people laugh. The scientist fails to see the humor, but can certainly understand the designer's intent. Later in the movie, Allen is sitting with Diane Keaton, playing a hip twenty-second-century poet wanna-be. In a romantic moment, she turns and asks him if he wants to have sex in the "orgasmatron." Allen answers that he doesn't need a machine to have sex, that his hands and other body parts will do. He is, after all, the maestro, jazz musician, manager of the Happy Carrot Health Food Store. Like the chattering teeth, then, the orgasmatron also has a clear function, even though it is not relevant to Allen's character.

At the root of our own knowledge of artifacts, particularly their function, is a more basic system, a set of core principles that enable us to recognize objects, classify them into different kinds, and predict their behavior. For example, we readily distinguish objects according to whether they are solid or not, whether they can move on their own or require an external force, whether they are man-made or natural, and whether they are living or non-living. How does the human infant acquire such knowledge? What about animals? More specifically, how do animals recognize objects, predict their behavior, and use this knowledge in the service of creating and using tools?

The argument I will develop is that all animals are equipped with a basic set of principles for recognizing objects and predicting their behavior. These principles form the core of their object knowledge, a simple yet powerful tool in their mental toolkit.

WHAT IS AN OBJECT?

When you pour yourself some coffee, how do you determine which object is sufficient for containing the liquid? When the coffee is poured into an object, such as a mug, what tells you that the liquid has been contained within the mug? The coffee has, after all, disappeared from sight. So what makes you believe that it is still present? If you watch Wayne Gretsky slapping a hockey puck, do you expect the puck to change shape? What could change its path to the goal?

These are the kinds of questions about objects that we don't consciously think about. We know that hockey pucks fail to travel through human beings because both objects are solid, and two solid objects can't be in the same place at the same time. We also know that coffee stays in the mug because mugs are nonporous containers. However, when we watch an ice hockey match or pour ourselves some coffee, we don't think about such explanations. We act automatically, allowing the innate knowledge of the brain to work as our guide. My claim that such knowledge is innate is based on studies of human infants and animals, as well as a set of intuitions from evolutionary theory.[2]

When organisms, including humans, encounter recurrent themes or statistical regularities, natural selection builds such information into their brains, making it an integral part of the survival system. What is innate in this situation is the mechanism for learning about a specific domain of knowledge, not the knowledge itself. Thus, the learning mechanism filters the experiences, guiding the organism to attend to some events in the environment, but not others. Returning to our earlier example, we are not born knowing about hockey pucks or coffee mugs. Rather, we are born with a learning mechanism that allows us to recognize objects such as hockey pucks and coffee mugs and make predictions about their behavior, whether they will move or stay in one place, fall apart or stay together, and so forth.

When a human infant sees a new toy, she reaches for it, guided by an almost magnetic force. Novelty is interesting and draws the infant's attention, as well as her hands. Now play a little trick. As the infant reaches toward the toy, place an opaque screen in her path. Not only will she drop her arms, but she will stop searching for the lost object. This finding, discovered in the 1950s by the famous developmental psychologist Jean Piaget, was used to argue that infants under the age of nine months lack *object permanence*—the capacity to represent an object in mind when it is no longer in view. You can do it: put this book down, place it under your chair, and imagine what it looks like. Easy, right? Absolutely. Using the reaching task and several other experimental manipulations, Piaget found that infants' knowledge of objects is impoverished before they reach their first birthday. As children grow older, they build on the experiences of seeing, touching, and moving to construct the domain of object knowledge.

Piaget's results on object permanence, and object knowledge more generally, have been challenged on theoretical and methodological grounds. If object knowledge (or any other domain of knowledge) is constructed in each child as a result of personal experiences, then each child should reach an understanding of objects at different developmental periods. Children growing up in cultures with frequent opportunities to interact with objects, especially objects with different behaviors, should acquire an understanding of objects at an earlier age than children given fewer opportunities. In contrast, if an innate mechanism is at work, then such varied experiences will have little effect on the core principles of object knowledge, even though American children may first learn about a hidden Beanie Baby and a child living among the Hadza hunter-gatherers in Tanzania may first learn about a hidden piece of fruit or meat. Universality is often a telltale sign of an innate mechanism at work.[3]

Cross-cultural analyses of child development reveal universality in the timing of object knowledge, particularly the acquisition of core principles. For example, most normal children begin reaching for hidden objects at around nine months, even though the range of their experiences with concealment is varied. Furthermore, infants' knowledge of objects (and as we shall learn in the later chapters, other domains) is far more sophisticated than what would be expected from the experiences they have encountered. This suggests that for human infants, the organization and acquisition of object knowledge is facilitated by a set of principles, innate mechanisms that guide the learning experience.[4] The same logic applies to animals.

Piaget designed his experiments to elicit explicit motor responses such as reaching. But if the child's motor system is immature, then reaching may not be a good measure of what she knows. To find out what human adults know, we typically ask questions. Although our linguistic responses may not always provide accurate representations of our thoughts, we generally assume a high degree of correspondence between what we know and what we say about our knowledge. Language is the mind's messenger. Unfortunately, we cannot use language to pose similar questions to infants or animals. Within the last fifteen years, researchers have met this methodological challenge by drawing on a technique that does not require reaching or language. Rather, it draws on the logic of magic and the simple fact that we look longer at events that intrigue us.[5]

When you watch a magician saw a body in half or cause a ten-ton box to levitate, you are drawn in, determined to figure out how the trick is pulled off. You know it is a trick, not because someone told you so, but because you have an understanding of objects. When you see a solid object levitate, your expectations have been violated. Solid objects can move in space only if they are launched, and they can remain stationary in the air only if they have an internal mechanism that allows them to do so. When we watch a humming-bird do the equivalent of Michael Jordan's pre–slam dunk space walk, we are not surprised because we know that this ability emerges from the speed with which the wings have been designed to move; even with his Air Nikes, Jordan can't maintain his body in one location in space for more than an eye blink. In contrast, when we see a magician move his arm and cause a chair, watermelon, or human to hang in space—with no strings attached—we stare. More generally, whenever we witness events that violate our expecta-tions, we focus our attention and do so because we are surprised. Events that are consistent with our expectations, however, enlist much less atten-tion. This is the crux of the research method used with infants, what those in the psychology trade call the "expectancy-violation procedure."

A newborn infant seeing a human levitate might find this event no more interesting than seeing a human walk on stage. A newborn might just men-tally register that it's normal for humans to walk on stage and float in space. If this is their worldview, then we would expect them to look no longer at levitating objects than at ones that lie stationary on a solid surface. In this sense, time spent looking at an object or event becomes a litmus test for a certain kind of knowledge. We look longer at the unusual. This is the essence of the expectancy-violation procedure.

In 1985 the developmental psychologist Renee Baillargeon designed an object permanence experiment for three- to four-month-old infants using the expectancy-violation procedure. Infants watched a solid screen rotate through an arc of 180 degrees until they were bored and looked away. Baillargeon then introduced a wooden block in the path of the rotating screen, and set up a curtain to hide both the block and the lower part of the screen. If infants are able to maintain a representation of the block in their mind, then when the screen rotates toward the block, they should expect it to stop at some point. In other words, although they can't see the block, they should expect the screen to make a rotation of less than 180 degrees; the screen should stop at the point of contact with the top of the block. This pattern is expected on the basis of a core principle of the object domain: one solid object cannot occupy the same space as another at the same time. In one condition, the screen stopped short of the full 180-degree rotation, whereas in a second condition—the magic trick—the screen rotated straight through the 180-degree arc. The magic trick represents a physical violation because the screen cannot move through the block.

If three- to four-month-old infants lack an understanding of object permanence, as Piaget claimed, then the magic trick isn't magic at all: once the curtain is in place, the block is no longer present in the infant's mind, and thus the screen is free to rotate through. In Baillargeon's experiment, therefore, a Piagetian would expect infants to look longer when the screen rotates less than 180 degrees because this represents a novel rotation. If, however, the infant has some understanding of object solidity and object permanence, then the 180-degree rotation is surprising because it violates a physical principle; infants are expected to look longer at violations than at events that are consistent with the physical world.

Infants look longer at the magic trick. Their eyes are telling us that they have detected a physical violation. More specifically, three- to four-month-old infants appear to understand that an object that is out of sight is still *in* the mind. Scientists have yet to determine whether this kind of knowledge about object permanence is different from what infants know by the age of nine months, when they begin reaching for hidden objects.[6]

Do animals keep objects in mind once they disappear out of sight? It is almost inconceivable to think that they don't. Imagine a group of gazelles foraging on the African plains. They have just spotted a lion lying low in the tall grass. The entire group becomes vigilant. The lion suddenly makes a move, darting out of view behind a bush. Would natural selection favor a mechanism that caused the gazelle to think, "Great, the lion has disappeared. Let's eat!"? Absolutely not. If such a mechanism evolved (*note*: not the language of thought that I have given to the gazelle, but a mechanism that failed to represent the lion's disappearance), it would clearly put the lion at a considerable advantage and ultimately cause the demise of the entire gazelle population. Predators are, however, locked into an ever escalating evolutionary arms race with their prey, each enjoying relatively brief periods of triumph. Evolutionary intuition, therefore, leads us to expect animals to have object permanence of some sort or another. The question is, what is the representation like in their mind? When a previously seen object disappears, do animals mark a location in space with something of general interest, something salient? Or does their representation preserve all the crucial properties of the object? Is a lion—a big yellowish cat with mane, tail, large paws, sharp teeth, that enjoys eating gazelles—concealed behind the bush or is the bush associated with something generally dangerous, something to avoid? Further, if the object disappears out of sight, do animals predict a trajectory, faithful to the kind of object detected and the mechanisms that underlie its capacity to move?

One of the first studies of object permanence in animals was conducted in the early twentieth century by the psychologist Edward Tinkelpaugh. An adult rhesus monkey watched as a banana was concealed under one of two cups. A screen was then raised, hiding the cups from the monkey's view. A few seconds later, the screen was removed and the monkey was allowed to choose a cup. The monkey consistently selected the cup concealing the banana. Tinkelpaugh repeated the same task with lettuce, a much less desirable food. Though the monkey appeared less interested in the lettuce, he nonetheless picked the correct cup. These studies show that when an object is placed out of view, rhesus monkeys maintain a representation of the object in mind, and this includes a depiction of the object's location. What is unclear, however, is how rhesus monkeys represent the object. They might, for example, think "something to eat under cup one" without storing any information about the kind of food concealed. Tinkelpaugh, however, ran an additional experiment to explore the possibility that their representation is more specific than "something to eat." The monkey sees a banana placed under a cup and then the screen is raised. While the monkey's view is obstructed, the experimenter removes the banana and replaces it with a piece of lettuce. When the experimenter removes the screen, the monkey chooses the cup previously containing a banana, but finds lettuce instead. Under these conditions, Tinkelpaugh reports that the monkey is surprised and angry. Although we cannot verify such impressionistic observations about the monkey's feelings, there is considerable appeal to the intuition that the monkey detected a violation based on what was stored in its memory. In particular, if we accept Tinkelpaugh's subjective observations, the monkey appears to have stored information about the kind of object concealed. He expected banana, not lettuce.[7]

How confident can we be about the monkey's representation, about the specific object it expected to find when lifting the cup? It may have represented something tasty as opposed to something that is edible but bland. Or something yellow as opposed to something green. Or something that is oblong with smooth curves as opposed to something that is ruffly and relatively flat. These alternatives must be explored before we conclude that the monkey was disappointed to find lettuce because it expected to find banana.

Research exploring how animal brains store representations in memory helps us understand whether the mechanisms underlying object permanence in animals and humans are similar or different. Here is a simple experiment. Show a rhesus monkey two empty cups, one red and one green. Place food under the red cup, but nothing under the green. Now give him a choice between cups. No-brainer: he readily picks the red cup. This task does little to tax the animal's memory because the objects remain in view; by simply keeping their eyes fixed on the red cup, the monkey can solve the problem. Now repeat the hiding game, but introduce an opaque screen in between the monkey and the cups. Keep the screen up for a few seconds, remove it, and allow the monkey to choose. Normal adult rhesus have no problem, rhesus infants fail, and adult rhesus fail if a specific part of their prefrontal cortex has been surgically removed. Apparently, this piece of the brain is necessary to store an active representation in mind—as opposed to something stored in long-term memory—and the brain has to be mature. These experiments help localize the general function of particular brain areas. They are deficient, however, in one important way: they fail to illuminate whether the prefrontal cortex (or some other brain area) represents a specific kind of hidden object or something more general and featurally amorphous.

We know, from hundreds of studies on rhesus monkeys, brain-damaged humans, and most recently, brain imaging (e.g., PET, fMRI) studies on normal humans, that processing of visual information depends on relatively specific regions of the brain, linked to different tasks or functions. These regions are roughly associated with detecting simple features (orientation, brightness, color), recognizing the identity of an object (what? which?), and determining the object's spatial coordinates (where?).[8] To determine whether the prefrontal cortex might integrate information from these visual areas of the brain, researchers ran Tinkelpaugh's experiment a second time, but this time with the aid of modern neuroscientific techniques that had been developed over a forty-year period. While the monkey watched an experimenter conceal and then reveal hidden objects, an electrode recorded the electrical activity of neurons in its prefrontal cortex. When there was a match between what was concealed and then revealed (e.g., banana-banana or lettuce-lettuce), the pattern of neural activity was different from that elicited by presentations involving a mismatch (e.g., banana-lettuce). The prefrontal cortex is, therefore, critically involved in storing representations of particular objects and using such representations to set up expectations about goals. As with Tinkelpaugh's original work, however, we don't yet know specifically how neurons in the prefrontal cortex code for particular features of an object. For instance, would we record the same patterns of neural firing if a yellow banana was concealed and then a brown, overly ripe banana discovered? Here we have a mismatch with respect to color, but not kind. It is still a banana. More work needs to be done.[9]

If an experimenter places food in a well and then covers the opening with a cloth, parrots, hamsters, cats, dogs, monkeys, and apes will go and retrieve the food. Their search behavior represents a simple understanding of Piagetian object permanence, that concealed food continues to exist. This ability appears to be present in most if not all animals. However, because they can solve this foraging problem by following the rule "move to the rewarded location," this experiment does not show that animals maintain a representation of the hidden food in mind. If we jump to the most difficult of Piaget's tasks, one that the subject can solve only by maintaining a representation of the food in mind, we begin to identify varying levels of ability among parrots, hamsters, cats, dogs, monkeys, and apes. Take the food and conceal it under a cup. Now slide this cup behind one of two opaque screens, remove the food, and show the subject that the cup is empty. Put the cup facedown. To find the food, the subject must infer that it was invisibly displaced and deposited behind one of the screens. Dogs, some monkeys, and apes successfully retrieve the food. Parrots, hamsters, and cats are clueless.

The differences and similarities between species on these tasks are intriguing, but there are interpretive problems. For those species that fail to show evidence of object permanence on reaching tasks, would we see evidence of this capacity using the expectancy-violation procedure? Are such animals similar to human infants who show some understanding of object permanence as revealed by their eyes but not by the action of their arms and hands? Furthermore, though animals may reach for objects placed out of sight, it is unclear how precisely they represent the object in memory. If a banana is hidden, do they store the representation of this object as "edible food," "yellow, boomerang-shaped object," or "reward"? As in studies of human infants, we need other tests to determine how animals represent objects.[10]

IN SEARCH OF CAUSALITY

Imagine a smoke-filled pool hall. Arranged on the green felt table are different colored balls. Your goal is to put one or more of the colored balls into a pocket. You start by taking a stick and lining it up with the cue ball, one colored ball, and a pocket. Next, you wind up and strike the cue ball. We can take this sequence of events and break it down into a series of cause-effect relations, each revealing the basic principles of your knowledge of objects. For example, you know that when the stick makes contact with the cue ball, it will cause a displacement if and only if the stroke has sufficient power; the stick will not, as you well know, pierce the ball or bend backward on contact. The pool stick, as a tool, represents the means by which you achieve the desired effect of causing a ball to move. As the colored ball departs, you immediately generate an expectation concerning its trajectory. You do not, for example, anticipate that the ball will jump over the others on the table or move through them; if they make contact with another ball along the way, they will be deflected off course. These basic principles are part of your object knowledge. You don't consciously think about such principles, but they nonetheless affect your behavior. Are these core principles part of the human infant's understanding of objects? What about animals?

Scientists have used the expectancy-violation procedure to explore how infants perceive causal interactions. The basic problem is whether infants would be surprised to see a colored pool ball spontaneously move toward the side pocket without being contacted by the cue ball or any other object. The answer is yes, they would. In fact, by the age of six months, infants appear surprised—look longer—when a stationary block of wood moves before it has been contacted by a moving block of wood (*contact* principle), when a ball rolls behind one screen and then emerges from a second screen without passing through the gap between the screens (*continuity* principle), and when one ball appears to travel along two different paths (*cohesion* principle). Infants are not, however, surprised when a stationary human moves away before being contacted by another moving human. The infant's eyes therefore reveal a certain sophistication about the principles of object contact and motion. Infants make the critical distinction between inanimate objects and animate ones like humans.[11]

Are infants as sophisticated about the principles of object motion when they must act on their knowledge by manipulating a tool or reaching for a target object? The answer is no. One-year-old infants are uncoordinated, but they can pull things that are within reach. To determine whether comprehension of physical causality is present early in development, the psychologist Peter Willatts sat infants next to the edge of a table with strips of towel placed out in front of them. On the distant surface of one towel sat a novel toy; the alternative towels were within reach, but other toys of the same kind were located off the surface. The task: pull the towel with a toy located on its surface since the towel is the only means to the toy. One-year-olds solve this problem immediately. This shows that they can represent the causal connection between an action on one object and its effect on another object. This comprehension is at the root of all tool use. These results do not, however, allow us to assess whether infants have a broader appreciation of which objects would bring the toy closer. For instance, do they understand that some changes to the towel are functionally relevant to the task where others are irrelevant? If we alter the color of the towel, the task remains the same. However, if we cut a strip out of the towel, creating a gap

All animals are equipped with three core principles for predicting the behavior of objects. Here, animals watch a film strip showing the core principle of continuity. When a ball passes behind one screen and then emerges behind a second, animals expect the ball to pass continuously through the gap separating each screen.

between the piece that can be pulled and the piece located under the toy, we have destroyed the towel's functional design.[12]

To understand the concept of a tool, one must minimally understand that certain features are functionally relevant whereas others are irrelevant. Although it is generally true that all washing machines and dishwashers are white, they could just as easily be rainbow-colored. Color simply plays no relevant functional role. By two to three years of age, children seem to know this. They ignore functionally irrelevant features such as color and attend to functionally relevant features such as shape, at least when it comes to such artifacts as tools.

Thus far I have argued that human infants are equipped with a set of core principles for recognizing objects and predicting their behavior. As core principles, they help direct the child to relevant experiences that ultimately shape her understanding of objects. When a child uses a tool, she reveals her knowledge of these principles, as well as her comprehension of object permanence and causal relationships. Animals use tools as well, but we know far less about the principles they bring to bear on this task.

NATURE'S HARDWARE STORE

When President Franklin Roosevelt was looking around for allies during World War II, British prime minister Winston Churchill stated, "Give us the tools, and we finish the job." We invent tools in response to environmental problems. We examine a problem, develop insights, and create just the right tool. But is this the only route to tool creation? No. Tools can be discovered accidentally or by deduction. Although deduction clearly requires some kind of smarts, including an understanding of if-then causal reasoning, the invention of many tools requires more, such as planning and insight. And sometimes we combine accidents with deduction to create a new tool.

Studies of wild animals do not enable us to determine with any confidence whether a particular tool user is the inventor. Nor is it possible to determine with accuracy the kind of mental calculations that went into the development of the tool. We simply observe an object that has particular design features and a user who wields the object in a particular way and in a particular context.

Consider what animals do with tools. For simplicity, I define a tool as an inanimate object that one uses or modifies in some way to cause a change in the environment, thereby facilitating one's achievement of the target goal. Most cases of tool use involve prefabricated objects, ones that are naturally provided by the environment such as a stone or a stick. In other cases an animal modifies an object, either by removing parts (e.g., stripping off leaves from a stick to facilitate insertion into a hole) or by combining two or more objects together (e.g., stone hammers and anvils); in these cases the modification improves its function.

The definition I have offered excludes using another individual's or one's own anatomy as a tool. For example, when the African honey guide (a bird) finds a honeycomb encased in a tree cavity, it grabs the attention of either a badger or a local human, leads them to the tree, and then recruits their help to open up the cavity and gain access to the honey. In this case, the honey-guide is using the badger and human in a tool-like fashion. The aye-aye, a nocturnal primate, uses its distinctively long, ET-like finger to tap on logs with potential insects. By tapping and listening, the aye-aye can determine which nooks and crannies have food. The aye-aye's finger is part of its standard equipment, but it is not a tool under our definition. Our definition also makes no reference to the kinds of thoughts or representations guiding tool use, and excludes general problem-solving situations from the pool of potential cases. For example, blue tits remove foil from milk bottles to skim cream from the surface, and ravens pull up pieces of string attached to meat in order to eat. Some genius tit and raven figured out a solution, but they did not use or modify an object along the way; the foil and string are part of the problem.[13]

The natural environments of animals are not like supermarkets or hardware stores. Though some foods are served, ready to eat, many desirable foods can be extracted only with a tool. One of nature's tricks is to wrap food in a hard, carry-around case. Fruits and vulnerable prey are often surrounded by a hard shell. Some predators crack this problem by using, and occasionally modifying, parts of the environment. Egyptian vultures approach the eggs of ground nesting birds with a stone, and like someone using a spoon to open up a soft-boiled egg, extract the goods with a calculated tap. Emulating a fighter pilot, the song thrush drops snails from a significant height so that they will crack open on the rocks below. While floating on their backs, sea otters use small stones placed on their stomachs to smash open mollusks. For the vulture, stones function as hammers, whereas for the thrush and otter, stones function as anvils. In none of these species, however, are hammer and anvil used in a coordinated fashion for the same task. In contrast, capuchin monkeys and chimpanzees place heavily armored fruits on an anvil and smash them open with a hammer; both stones and heavy branches are used as hammers and anvils. In no case of stone tool use is there evidence that individuals modify the structure of the stone in the service of improving the tool's functionality, though a few cases of anvil stabilization have been observed; for example, in 1996 the psychologist Tetsuro Matsuzawa observed four cases of wild chimpanzees placing a small rock beneath a larger one to level the smashing surface of the anvil.[14]

Research on stone tool use raises an important yet general question about the mind of the tool user. Is the choice of hammer or anvil a random process? Does, for example, the thrush pick out a location where the odds of hitting a good solid rock are high, rejecting places where there is a scarcity of rocks? In finding a good anvil, do animals optimize the surface features of the stone, its curvature, and so on? To date, the only useful information on this problem comes from some preliminary field experiments and observations on chimpanzees. In particular, if stones are placed in a pile, chimpanzees appear to pick out those that meet some minimal criteria for functionality: flat, small enough to be held in the hand, hard enough to crack open a nut. Having found an appropriate hammer, chimpanzees in the Tai Forest of the Ivory Coast leave their hammers near a fruiting palm nut tree, returning at some point to use them again. This suggests they understand the function of these stones and appreciate the importance of retrieving them in the future. Young chimpanzees require several years to understand what, precisely, constitutes a functional hammer and anvil set.[15] It is not yet clear whether chimpanzees have an understanding of optimal design when selecting a hammer and anvil, and whether they rank the relevant as opposed to irrelevant features of these tools. One study of wild crows and three laboratory studies of primates provide us with a set of insights into this more complex problem.

In general, most of the tools that animals use involve the implementation of gifts from nature: large leaves used by chimpanzees as waterproof seats for the moist ground, by orangutans as protection against the hot sun, and by bonobos as rain hats; dry pods used by vervet monkeys to extract sap from a tree; twigs used by green herons as bait for fish. When natural objects are modified, the craft is subtle: to extract ants and termites from their homes, chimpanzees strip the leaves off a stick, insert the smaller end into the hole, and wait patiently for their victims to climb up and out; to obtain water or sap from the holes in a tree, chimpanzees chew leaves to create an absorbent sponge. So subtle are these tools, especially when contrasted with something as simple as a hand ax from our early ancestors, that if you stumbled upon them in the forest, you probably wouldn't recognize them as tools. The lack of design is striking. However, some fascinating observations of crows from New Caledonia suggest a small step up in the ability to manipulate an inanimate object for a specific function. New Caledonian crows construct two types of tools, each with a highly standardized design. The hooked twig type is long, thin, and curved, and can readily be moved into a narrow hole in a tree to extract insects. The stepped-cut twig is much wider, thicker, and barbed, facilitating extraction of insects from difficult holes. These designs are not random, but rather, customized with respect to the extraction problem.[16] Although the crows create tools with specific design specs, we still don't know how they represent such tools. We don't know, for example, whether they evaluate potential tools with respect to functionally relevant and irrelevant features. If we painted a tool with the right curvature pink, orange, and purple, would they realize that such color changes are irrelevant? And if the potential tool was shaped in the right way but constructed of a flimsy material, one that snapped when inserted into a hole, would they realize that material is sometimes a more important feature than shape?

Most studies of tool use in animals focus on cases where the animal uses an object to gain access to food. One of the most fascinating cases of tool use in a nonfood context comes, however, from observations of chimpanzees using sticks for dental checkups. Grooming with hands and mouth is common in many primates and plays an important role in establishing and maintaining social relationships. It is, therefore, a friendly gesture. In 1973 the primatologists William McGrew and Caroline Tutin observed several cases of dental grooming where one chimpanzee used or modified an available stick and then began working on the patient's teeth. In describing the two most common partners in this captive group, Belle the dentist and Bandit the patient, they note that

> the usual procedure was to begin normal social grooming directed to
> the hair of his limbs, torso, or head. She then concentrated on his face

and, finally, opened his mouth. If he cooperated by gaping his jaws, she began dental grooming with the fingers. . . . Sticks used as supplementary aids were picked up from the surrounding substrate. . . . [In one incident] Belle picked up a twig approximately 0. 5 cm in diameter and stripped off the leaves, leaving a pencil-like object about 15 cm long. Between sessions of using it as a pick, she left it hanging loosely out of the corner of Bandit's mouth.

The chimpanzees' detailed focus and goal-directedness is fascinating. More important, however, is the fact that in chimpanzees, tool manufacture and use represent a response to a broad range of problems. As McGrew and Tutin suggest, such dentistry appears to have emerged in response to the fact that Bandit, a young male, was just beginning to shed his deciduous teeth and also appeared to be walking around with a toothache. Although the tools used for grooming were few, they all consisted of functional design features, including scraping edges and pointed ends, useful for picking between the teeth.

Capuchin monkeys are unquestionably the most dexterous of the New World monkeys, and relative to their size, have the largest brains as well. Over the past ten years the primatologist Elisabetta Visalberghi has designed a series of experiments to explore their technical sophistication. The basic approach involves presenting capuchins with a potential set of tools and a problem, such as a peanut lodged out of reach in a transparent tube. To solve the problem, the capuchin must create a novel tool by combining two or more objects from the set. For example, to reach the peanut, the capuchin might insert two sticks into the tube, one after the other. With one stick in place behind the peanut, the other can be used to push. Some individuals quickly solve this problem, ignoring objects that are functionally useless; for example, although a piece of rope would fit inside the tube, its lack of rigidity makes it an ineffectual tool. Other subjects, who had solved the problem on earlier tries, occasionally made some remarkably comical errors, trying to fit a thick stick inside a smaller tube. Similarly, individuals who learned to smash open nuts with stones and extract the food with a twig sometimes reversed the process, with a hilarious lack of success.[17]

These experiments show that in thinking about a functional tool, capuchin monkeys attend to some of the relevant features. They appear to understand that some objects are potential tools and others are not. But their understanding is incomplete, as evidenced by the errors they make. We therefore need a better understanding of whether they ignore featural changes that are irrelevant to a tool's functionality and rank tools according to something like an optimized feature set.

In 1999 tamarins in my lab were presented with two dark blue canes placed on a light blue tray, separated by a barrier. A white food pellet was located on the inside of one cane's hook, and on the outside of the other cane's hook. The only way to obtain the food was to grab and then pull the shaft of the cane; this action advanced the pellet and allowed the tamarin to reach out and grab it. Although either cane could be used, the best cane (given a single choice) was the one with food located inside the hook; by simply pulling this cane straight back, the tamarin caused the food to advance.

The tamarins readily solved this problem, pulling the cane with a pellet located inside the hook. They could, however, have solved the problem without understanding that the cane is a tool, and that it is only one of a range of possible tools. For example, perhaps they learned to pull a blue object to attain a white piece of food. If this is all they learned—a rather inflexible level of comprehension—then changes in the color, shape, texture, or dimensions of the potential tool and food would aversely affect their performance. When these changes were imposed, the tamarins' performance revealed two interesting patterns. First, they did as well with new tools as they did with the original blue canes, even though the new tools were radically different. Second, and most important, they preferred to use new tools with functionally irrelevant changes as opposed to tools with functionally relevant changes that made the task more difficult. Thus, for example, if the color of the cane changes from blue to red, but shape, size, and material stay the same, then the tool is still perfectly functional. A color change is functionally irrelevant. In contrast, if the hooked portion of the cane is on the bottom and the straight edge is on top, closest to the food, this relative change in shape is functionally relevant; it is much more difficult to retrieve the pellet with a straight edge than a hooked tip. These results indicate that tamarins attend to the relevant features when picking a tool. Their choices make sense given the task.[18]

Perhaps the most stunning demonstration of cause-effect comprehension in the domain of tools comes from the psychologist David Premack's work with Sarah, his star chimpanzee. In the 1970s Premack presented Sarah with two slides showing a before-and-after sequence such as an apple–two half apples, dry sponge–wet sponge, and blank paper–written-on paper; some sequences were familiar to Sarah, some were not. Next, Sarah was presented with three utensils. Only one utensil was appropriate with respect to the transformation. Thus, for example, when a subject is shown apple–two half apples, knife is the appropriate tool out of a set that also includes pencil and bowl of water. Sarah consistently selected the appropriate tool, knife for the apple pairing, bowl of water for the sponge pairing, and pencil for the paper pairing. Does this show that Sarah understands the causal relationship between tool and transformation? Not quite. It is possible, for example, that Sarah has merely learned that apples are often associated with knives, pens with paper, and water with sponges. If this is correct, then Sarah does not understand that one object plays a causal role in transforming another object. She doesn't understand, for example, that knives, but not bowls of water, can transform one apple into two pieces.

To distinguish between these explanations, Premack went further with his experiments, and so did Sarah. The task involved the completion of a pseudo-*sentence*, a kind of fill-in-the-blank test. For example, when the sequence is apple–two half apples (before-after), knife is the appropriate choice out of the set including tape and bowl of water. However, when the sequence is two half apples–apple (after-before), tape is the correct choice out of the same tool set. Sarah succeeded on this task, with many different object pairings, including novel ones. This shows that she understands more than the associations between objects. She understands that some objects are designed with specific functional roles, transforming other objects into alternative states.

Crows, monkeys, and apes have some understanding of the functional properties of tools. Not only are they good tool users, they also understand that for a tool to be useful, it must have certain design characteristics. In contrast to our early human ancestors, however, animals have a limited capacity for refining and combining objects to make better tools. Having experienced the creation of even a simple tool—a crude stick for performing a dental operation or a branch for walking over the kapok's thorns—animals don't go on to make better dental tools or Birkenstock sandals.

A MATTER OF DESIGN

There is a joke that goes something like this. A man is conducting a survey to find out what people consider the best invention of all time. After hearing peoples' votes for the telephone, airplane, computer, transistor radio, and such, he encounters a distinguished old gentleman who responds, "The thermos." Stunned, the surveyor asks, "If you don't mind my asking, good sir, your answer is unique among the million or so answers I have received. Why do you think the thermos is such an extraordinary invention?" Leaning back in his rocking chair, the gentleman responds, "Well, you see, it keeps cold things cold and hot things hot. How does it know?"

Of course, the thermos doesn't know. But as a tool, it has been designed to maintain the temperature of the substance stored, no questions asked. All tools share the property of design, and so do living organisms. But how do we identify something with design? As the cognitive scientist Steven Pinker puts it in his 1997 book *How the Mind Works*, things with design share several crucial properties, including "signs of precision, complexity, efficiency, reliability, and specialization in solving its assigned problem, especially in comparison with the vast number of alternative designs."[19] Whereas natural selection is responsible for an animal's design features, the animal mind is responsible for a tool's design features. Animals from almost all walks, flights, and swims of life use tools, but to understand the mind that creates tools, we must push our theoretical analysis much further.

In commenting on chimpanzee tools, particularly the similarities and differences between chimpanzee and human tools, the archaeologist Steven Mithen makes four points in his 1996 book *A Prehistory of the Mind*. First, the design of the chimpanzee tool is simple with respect to the number of parts or modified components. Compare the chimpanzee's termite fishing stick with a human's fishing pole, one with a reel, line system, specialized hooks, and weights. Second, the chimpanzee manufactures the tool using the same hand and arm movements as for other behaviors. The actions we use to carve a spear or craft a bowl on a potter's wheel are unique gestures. Third, the tools chimpanzees create have a narrow range of functions, and there is little evidence that the creators can think up new functions for the same tool. Humans can use a knife to cut meat, hold down a stack of papers on a windy day, pry open a lid from a jar, or dig dirt out of a small crack. Last, when a new tool is invented, other chimpanzees are slow to pick up its functionality. In humans, new tools, especially useful ones, spread through the population almost immediately; the rapid spread is largely due to our capacity for imitation, a capacity that may be lacking in all other animals. These attributes, concludes Mithen, indicate that unlike humans, chimpanzees lack a "technical intelligence devoted to manipulating and transforming physical objects."[20] Everything that I have discussed in this chapter supports the first three points. The fourth point, which we haven't yet discussed, will be addressed in chapter 6.

Should we accept Mithen's claim that chimpanzees and other animals lack a technical intelligence, one that our earliest ancestors evolved about two to three million years ago? Perhaps. I think, however, that this is the wrong question. After all, human infants also lack a technical intelligence of the sort Mithen envisages. I would suggest a more interesting set of questions. What constrains the development and evolution of a technical intelligence? Are the deficits that animals and human children exhibit due to an impoverished set of core principles in the domain of object knowledge? Are they due to a lack of dexterity? Patience? Creativity? Or are human infants and animals technologically challenged because their problem-solving abilities are impoverished when it comes to inanimate objects? These questions are fundamental to our understanding of technical sophistication, and bring us full circle back to the notion of design.

In 1984 one of the vervet monkeys that I observed in Kenya acquired a fascination for the carburetor in my jeep, frequently coming over for an inspection. Now, this fascination certainty did not result in a mechanic's degree. And although I never administered a test, I don't think he could discriminate carburetors from other engine parts. How might he acquire such knowledge? Carburetors are what they are because of their function, because of the role they play in engines. For those of us who are technically challenged, we learn about such parts by instruction; someone tells us that carburetors were designed to combine just the right mixture of air and gas. We learn of the part's intentional history, the designer's plans and goals.

Some cognitive scientists such as Paul Bloom believe that having a representation of an artifact requires an understanding of what the object was designed for. Consider the chair, something far simpler than a carburetor. A feature-by-feature breakdown of chairs reveals little useful information. All chairs have four legs, right? Wrong. Imagine two square pieces of wood connected at a ninety-degree angle and supported by a cylinder. No legs. What determines whether or not something is a chair is whether the designer created it for sitting. Even if the design is poor, we still call it a chair! Artifacts are defined by their function, in the same way that our body parts are defined by function. Young children seem to understand this, as evidenced by the kinds of questions they ask about objects in the world. When a child asks, "What is this for?" she is usually looking at a novel artifact or some *part* of a living thing. As the developmental psychologist Frank Keil has documented, children never pick up a rock or some sand and ask, "What is this for?" Give them a screwdriver or a pressure gauge, however, and off they go. Similarly, children never ask this question about a cat or a bush, but they will ask about the cat's ears and tail, and about the thorns and roots of a rosebush. In the words of Daniel Dennett, the *design stance* that children and adults take toward objects depends critically on the kind of object confronted.[21]

Paul Bloom provides an illustrative example. Your watch is broken. The minute hand no longer advances. You bring it to the watch store on Monday. On Tuesday your watch is sent out to the factory and disassembled, minute and hour hands in one place, crystal face in another, internal motor in yet another. Is it a watch on Tuesday? It is certainly not a functional one, but it is a watch, a broken watch. On Wednesday it returns, intact and functioning, telling accurate time. Artifacts can be broken and retain their status because we recognize the designer's intent. In fact, even if the designer builds something that isn't remotely functional—it has bad design—it retains its status by virtue of the fact that there is an intentional history.[22]

If understanding the intentional history of an artifact is critical to creating the representation, then unless an animal witnesses the invention, it presumably has no hope of acquiring such a representation. The same holds true for an adult human. Even the most eloquent signing apes (see chapter 8) can't ask about intentions and receive a comprehensible answer. And neither can human infants. We are thus left with a dilemma. If we are interested in the extent to which we share common representational ground with animals, particularly their representation of artifacts, then we need to distinguish between the enhancing properties of language and the necessity of language in developing a concept of an artifact. Although our work on tool use and manufacture is in a relatively primitive state with respect to understanding the minds of the users and makers, my own hunch is that language enriches the human representation of an artifact, but is not necessary. For all organisms, the representation of an artifact grows out of a more general knowledge of objects. In this sense, the mental tool that allows all animals, humans included, to recognize objects and predict their behavior is at the core of their representation of artifacts. At present, we don't understand how language transforms the child's representation of artifacts or enriches the adult's representation. One insight into how this transformation arises comes from looking at a second tool in the mental toolkit, the ability to enumerate and understand the abstract notion of number.

The second installment in the universal toolkit is a mental tool that allows all animals to assess the number of objects or events, be they seeds, bananas, pacifiers, or coins.

3

Number Juggling

Bertrand Russell, the great British philosopher of the early twentieth century, observed that "It must have required many ages to discover that a brace of pheasants and a couple of days were both instances of the number two."[1] Perhaps this insight into the complexity of the number system explains some old birding wisdom. To watch most birds without disturbing them, one needs to hide behind a blind, a kind of tent. If the bird sees you enter, however, you have gained little because it is now aware of the blind. One way around this problem is for two people to enter the blind together and then, some time later, for one person to leave. Upon seeing the departure, the bird apparently assumes that the coast is clear, and goes back to business as usual. Why? Because most birds observed in this situation are incapable of computing a simple subtraction: $2 - 1 = 1$! If true, this does not say much for the mathematical prowess of birds, to say nothing of other animals who might be similarly deceived.

Animals living under natural conditions must have some numerical abilities as evidenced by their feeding behavior. We know, for example, that in species with widely different life cycles (flies that live for a day, tortoises that live for a hundred years or more), ecologies (rain forest, desert, savanna), and mating systems (monogamy, polygamy, polyandry), individuals engage in a form of mental calculus designed to maximize energetic intake on foraging trips. A wide variety of animals calculate average rates of return in one patch of food so that they can either stay or switch patches; use information about search costs and search speed to assess optimum rates of energetic returns; obey Bayes theorem, a calculation of the probability of future returns based on prior experience, to make sensible decisions about where to feed next; hide thousands of seeds over a broad swath of turf and then return months later to retrieve most of their stash.[2] These calculations show that animals are equipped with powerful number-crunching devices, but they say little about how animals represent number in their head.

In all social animals, competition within and between groups is typically guided by the dictum "there is strength in numbers." In chimpanzees, attacking and killing a member of another community occurs only if the intruder is alone and there are at least three adult males in the attacking party. Within social groups, individuals also form coalitions of two or more members to increase their relative dominance over a third individual. In bottlenose dolphins, such coalitions can reach exceptional levels of sophistication. Two to three males in one coalition join up with a second coalition to defeat a third. Occasionally a large group of fourteen dolphins forms a team that readily overpowers smaller groups. And for what? A single, sexually receptive female. What we don't know in this instance is whether a group's superiority is due to overall numerical superiority or to something else. For example, what if the total number of individuals in the two united coalitions is four and the number of individuals in the single coalition is five? Here, two coalitions are greater than one, but five is greater than four. If dolphins are truly counting the number of individuals, then such differences matter.[3] Studies of dolphins have yet to shed light on these questions. However, studies of a land-based mammal—the African lion—provide some answers.

Lion prides consist of up to eighteen adult females and their offspring, accompanied by one or two adult males. One threat to a pride is the possibility of a foreign male takeover. When new males enter a pride they often kill whatever cubs are present. Such attacks are designed to speed up females' sexual readiness and receptivity to a male's amorous advances. Other lion prides also pose a threat because they will fight over territorial resources. To determine whether a foreign group poses a threat, lions would need to determine the number of competitors by means of their vocalizations. That is, listen to the number of lions roaring at a distance, and assess whether they are numerically superior or inferior.

In 1994 the biologists Karen McComb, Craig Packer, and Anne Pusey conducted playback experiments on the Serengeti plains of Tanzania to determine whether lions compute the number of individuals in a distant group on the basis of their roars alone. Lions, like many other species, have individually distinctive voices. Thus, when a roar is heard, listeners can determine whether it is a familiar individual or a foreigner. Two different tapes were constructed, one with a single unfamiliar female roaring and one with three unfamiliar females roaring. These tapes were then played back to prides differing in size and composition. When roars of one intruder were played, members of the pride were more likely to approach than when roars of three intruders were played. Furthermore, when pride size was small, individuals tended to roar to increase the odds that others would help in defense.

What do these results tell us about the numerical abilities of lions? It is clear that the lions hear a difference between the two tapes because they respond in different ways. But is the key difference between these tapes the number of callers? Maybe, maybe not. Although one tape consisted of one caller and the other consisted of three, there were other differences as well. For example, one tape played for longer than the other and consisted of a greater variety of frequencies and intensity changes; in choruses, calls overlap and there is a complex interaction between the frequency range and loudness of each caller. Such factors cloud the interpretation. Lions may be able to discriminate vocal signals on the basis of the number of callers or may, more simply, discriminate such signals on the basis of their overall duration or intensity. To distinguish between these possibilities one might play back a signal with one lion roaring and a second signal with the same lion roaring three times in a row. In this case there is one roar versus three, but only one caller.

Even if lions have a capacity for numerical assessment, and my own hunch is that they may, it is not yet possible to determine what is going on in the mind of a lion during these simulated threats. Lions might be counting the precise number of individuals (e.g., one versus three callers) or they might be doing something more rudimentary, such as contrasting "one" versus "many" callers. To understand a lion's numerical abilities, we would need to run tests involving one, two, three, four, and more roaring lions, and for each test measure something like the intensity with which they respond. If our only response measure is approach or avoid, then we cannot test the proposed hypothesis. For example, they might approach roars of one or two lions, but avoid roars from three or four. This shows they discriminate numbers less than three from three or more, but doesn't show whether they discriminate one and two, or three and four. In contrast, if we can measure the speed with which they approach or avoid, or the number of roars elicited, then we can. If lions do count "one, two, three, four, . . . " then when they hear one individual roaring they should respond more intensely (aggressively) than when they hear two, and their response to two individuals should be greater than it is to three, and so forth. In contrast, if lions merely compute "one, more than one," then we will find a difference between one roaring lion and all other numbers greater than one, but no difference between two, three, and four.[4]

Studies of lion roars, together with comparable experiments on chimpanzee food calls,[5] indicate that animals spontaneously, and without training, exhibit rudimentary numerical abilities. Although lions may count the number of individuals in a roaring chorus, and chimpanzees may count the number of food items discovered in a patch, an alternative interpretation is that these species make crude quantity distinctions, corresponding in a loose way to our concepts of "few," "many," "one," and "more." If the latter interpretation is correct, other questions emerge. What social or ecological pressures would favor a mind capable of more precise numerical quantification? Are there conditions under which more sophisticated numerical capacities can be elicited? In tackling these questions, we need to understand what is at stake. The best way to do that is to show how the minds of human adults represent number, and then explore how young children acquire this mental tool.

WHAT'S IN A NUMBER?

We count objects and events all the time. Some counting requires our complete attention, as when we change foreign money at a bank or put coins into a parking meter. In other cases, we obtain information about number through an effortless process, as when we see dice on a board game. There is no need to count the number of dots. The total pops out because of a particular pattern such as "snake eyes" or "boxcars." When we count, and are aware of it, at least three computational abilities are called into action. First, we focus on the objects or events to be counted. Second, we identify cases of the relevant class of items or events. Our visual system has an exquisite ability to detect contours, boundaries, and episodes of motion. Such information lies at the foundation of our capacity to individuate objects or actions. Third, we establish a system for labeling each item in the count list with a symbol. For most adults in the world, counting is a cinch. We do it as effort-lessly as eating. Moreover, in all the world's languages, there are words for identifying number, for marking quantity. But for human infants, young children, and animals, counting represents a challenge—one that may derive from problems of attention, object discrimination, or symbolic pro-cessing. Although human children ultimately grow into number jugglers—investment bankers, tax attorneys, and computer programmers—their understanding of numbers takes years to develop. And for animals, learning the most elementary properties of the number system is difficult.

What cognitive abilities are involved in counting? We must first appreciate that some forms of numerical assessment look suspiciously like counting, but are not. When we view a small set of items, we often determine the number in the set instantaneously. We don't recall counting off the items in the set, but we can nonetheless state the total. If someone drops three coins on a table, we are able to determine the number without counting each coin. This process, which allows for rapid numerical quantification, is called *subitization*. Some researchers interested in how our brains process the visually presented world believe that it is a fundamental, unconscious sys-tem for recognizing patterns.[6]

Subitization is possible within the range of three to four items. Above this range, we must count. And to count, we must understand five core prin-ciples. First, there is a *one-to-one correspondence* between the items to be counted and the labels we apply to them; each label or symbol is unique for the item just counted. Thus, if we see six cookies, we don't count, "one, two, one, three, two, three," but rather "one, two, three, four, five, six" or "moja, mbili, tatu, nne, tano, sita" in Swahili. Second, the order in which labels are applied is nonrandom, constrained by the formality of *ordinality*: mathe-matics requires a stable sequence. In counting cookies, we can count each one "one, two, three, four, five, six" or we can arrange cookies into groups, and then count "two, four, six." Third, any solid object or discrete action can be counted—the principle of *property indifference*. Thus, we can count stars in the sky, gazelles in a herd, pieces of spaghetti on a plate, the number of times a tennis ball sails across the net, and the number of times an army lieutenant does jumping jacks. Nonsolid objects such as sand, water, mer-cury, and pudding cannot be counted. We may ask for two bowls of pudding, but we don't ask for two puddings. The distinction between solid and non-solid objects shows the intimate relationship between number and object knowledge—a connection between two mental tools. Fourth, we can count items in any order—the principle of *order indifference*. If we are interested in the total number of cookies on a plate, we can count the chocolate ones first and the oatmeal ones second, or the reverse. Finally, the last label applied in a count sequence represents the total number of items counted—the principle of *cardinality*. While counting cookies, we can count "one, two, three, four, five, six," "two, four, six," or "one, three, six." The number "six" is the cardinal term. It indicates how many items are in the attended set.[7]

When a human states that there are seven objects, what have we learned about his understanding of number? To answer this question, we need to draw a distinction between a *category* and a *concept*. A number category, like the category *solid objects* or *verbs*, is a category by virtue of the fact that it refers to specific things on the basis of their properties. In the case of number, the essential property is the countable item, action, or event, independent of its physical attributes. Thus, the mental category of "seven" can refer to seven penguins, seven sins, seven wonders of the world, seven days, seven computer beeps, seven odors, or seven licks on an ice cream cone. In contrast, a number concept represents a symbol that has a particular relationship to other symbols within the number domain. Like nouns and verbs that hold a particular relationship to each other in the structure of a sentence—the domain of grammar—number concepts have unique roles by virtue of the arithmetical operations that can be performed on them and with them. Thus, the concept of "seven" is unique because it is a prime number, the sum of one plus six, the number of cookies left when we subtract two from a stack of nine, and the only integer that is less than eight and greater than six. The category-concept distinction is important because a human child pointing to a pile of cookies and saying "seven" may know that there are seven cookies, but may not know that this represents one less than a pile of eight cookies, six more than one cookie, and so forth. One might well imagine, in fact, that children acquire a number category well before they acquire a number concept. Similarly, some animals may have evolved a number category, but not a number concept.

When do children start counting? We know, from several experimental studies, that some of the core principles underlying counting are in place well before children have acquired language.[8] Specifically, infants can discriminate sets of objects on the basis of their numerical differences and understand the one-to-one correspondence between numbers and objects. But actual counting with numbers does not start until the second year of life, and then takes almost two more years before the child fully grasps the five core principles of counting. How do we know this?

Here are two simple tasks. If you ask the typical two-year-old to hand over five cookies out of a plate of ten, she will grab a handful and pass them over. If you tell her that the bunch she handed over is not five, and ask her to try again, she will either rearrange the cookies you have or grab another bunch. Though she seems to understand that you want *some* cookies, she is not counting. By the age of four, the child will count and hand over precisely five cookies when asked to do so. Now try a different task. Instead of asking for some of the cookies, simply ask the child to count how many there are on the plate. After she answers, ask her again. If she understands that the final count term (i.e., the cardinal value) is the correct amount when she has counted correctly, on the second query she should respond with the final count term alone—there is no need to count a second time. Two-year-olds count over again and again and again. Three-and-a-half- to four-year-olds answer with the final count term. The older children understand cardinality—the idea that the last count term signifies the total number counted—while younger children do not.

Children learn mathematics by tapping into the fundamental building blocks of counting, starting with simple addition and subtraction—noticing variations in quantity. Much of this knowledge builds on the mental tools from the object domain. Then, through tutoring, our swiftness with numbers progresses to (in some people at least) the dizzying heights of calculus, Euclidean geometry, and set theory. But is there any evidence that children under the age of four years understand some of the more basic mathematical operations? What about infants without language?

Imagine that you are in line at the supermarket and you glance over and read the following tabloid headline: "HUMAN BABIES BORN DOING ADDITION AND SUBTRACTION." A skeptical reaction would be understandable, but unwarranted. Unlike other tabloid headlines concerning space aliens, human babies weighing a ton, and Hollywood smut, this one is for real.

Recall that when an object is placed out of sight behind a screen, human infants expect the same object to be in the same place when the screen is removed—they have *object permanence*. But does the infant expect only one object? Would she be surprised to find two? Taking advantage of the human infant's capacity for understanding object permanence, the developmental psychologist Karen Wynn ran an experiment in 1992 to explore whether five-month-olds can solve addition and subtraction problems. Using the expectancy-violation procedure discussed in chapter 2, an experimenter first familiarizes the infant with the key objects and nonmagic events to remove the effects of novelty. In this particular study, an experimenter shows an infant either one, two, or three Mickey Mouse dolls on a stage, as well as a screen moving up and down. Test trials start once the infant is bored, looking away from the stage. In the "expected" test $(1 + 1 = 2)$, an infant watches as an experimenter lowers one Mickey Mouse doll onto an empty stage. A screen is then placed in front of Mickey. The experimenter then produces a second Mickey Mouse doll and places it behind the screen. With the screen removed, the infant sees two Mickeys on stage, an outcome that should be expected. No magic. In the "unexpected" test, the infant watches the same sequence of actions, involving the same two Mickey Mouse dolls, but with one crucial change—a bit of backstage magic. When the experimenter removes the screen, the infant sees either one Mickey (i.e., $1 + 1 = 1$) or three (i.e., $1 + 1 = 3$). Although their jaws don't drop in surprise, five-month-olds consistently look longer when the outcome is one or three Mickeys than when the outcome is two. And precisely the same kind of result emerges from an experiment involving subtraction instead of addition.

Wynn concluded that infants have an innate capacity to do simple arithmetic. By simple, she meant addition and subtraction with a small number of objects. By innate, she meant that the general capacity to track objects and perform mathematical operations on them comes as part of our standard genetic equipment.[9]

Wynn's results raised fundamental questions about the development and evolution of nonlinguistic representations. For example, what kind of representation does an infant have while watching Mickeys come and go? Do infants have access to nonlinguistic mental symbols—what the cognitive scientists Randy Gallistel and Rochel Gelman call "numerons"—that allow them to tag individual objects as they appear and then disappear behind a screen? When an infant sees one Mickey, does a special symbol light up in her head? When a second Mickey is introduced, does a different symbol light up? Or maybe they lack such integer symbols altogether. If so, they might

Three possible mechanisms for assessing number in human and nonhuman animals. In this illustration, the human infant represents number by a symbolic integer system that provides a one-to-one correspondence between the number of objects and a specific symbol. The rat represents number in the same way that a graduated cylinder registers the volume of liquid poured, converting the amount of liquid into a digital number. The chimpanzee represents number as an object file, with each file corresponding to a slot in memory.

assess the number of objects seen by storing each one in a memory slot, a folder in the brain's filing cabinets that can be retrieved and processed at some time in the future. Alternatively, perhaps they compute number by means of an internal metronome, a ticker that records the number of objects perceived. As each object is registered, a record accumulates in much the same way that mercury registers changes in temperature or sand fills up a graduated cylinder. These are all possible mechanisms for computing number in human infants. And they are possible mechanisms for animals as well.

Human infants begin life with a rudimentary numerical capacity. Over the course of approximately four years, they further develop this capacity and do so without extensive teaching by adults. The counting system emerges spontaneously, but is guided and constrained by a core set of principles: one-to-one correspondence, ordinality, property indifference, order indifference, and cardinality. With time, formal schooling takes this raw material and turns some children into math wizards and others into members of society who can understand averages and percentages.

COUNTING SHEEP

In the early twentieth century a former math teacher, Willhelm von Osten, claimed that he had discovered a horse named Hans with exceptional talents. Hans used hoof taps to communicate, with each hoof tap pattern corresponding to a number or letter. When Hans was on stage, he was presented with a question, either spoken in German or written on the blackboard. Of interest to us here is the fact that some of the questions involved mathematics, in particular, problems of integer addition and subtraction as well as comprehension of fractions. Hans readily solved such problems, both when von Osten was present and, to a large extent, when he was absent.

Von Osten invited skeptical researchers to scrutinize his horse's abilities. If Hans's virtuoso performance was real, then his math skills were at least as good as those of many teenagers with schooling. Consider the evidence. Hans is on stage and either hears a spoken question or sees a problem on the blackboard. To decipher the question, he must be capable of either breaking down the sound of spoken German into meaningful words and phrases or translating symbols on a blackboard into something meaningful. Either, in and of itself, would be an exceptional capacity.

Let's assume that Hans has the capacity to extract words and decipher the meaning of a scribble on the blackboard. Now he must take those symbols and run the relevant computation. How might Hans have solved such problems? Although the "Hans Commission," composed of an impressive group of scientists, originally concluded that he must legitimately be doing math, a scientist by the name of Oskar Pfungst demonstrated, beyond a shadow of a doubt, that Hans was no mathematical savant. It turned out that Hans's ability to do math depended on the presence of at least one person who knew the answer to the proposed question. If no one knew the answer, Hans flopped. Given this clue, what allowed Hans to pass with a knowledgeable audience and caused him to fail with an ignorant one? The answer is not telepathy, but something much simpler.

The human body and face convey information. Hans was clever enough to discern it and use it. By observing his audience members, Hans extracted information about when to stop tapping. Although no audience member was aware of their communication, they provided Hans with the answers. In fact, even after Pfungst figured this out, he was incapable of withholding the necessary cues from Hans—some part of Pfungst's body, such as his eyebrows, always signaled Hans to stop tapping at the appropriate moment. Hans had no understanding of number or mathematics, but he was brilliant with respect to decoding body language, and human body language at that.[10]

In the mid-twentieth century psychologists showed that they could train animals to do amazing things. By rewarding some actions and punishing others, they could turn Fido the dog into a pirouetting Nureyev, and a rat into Glenn Gould at the piano. But could animals trained on some of the basics of the number system spontaneously develop an understanding of number that was both abstract and consistent with some of the core number principles described earlier: one-to-one correspondence, property indifference, cardinality? And if the answer is yes, can we assume that the representation of number in animals and humans is the same? Consider a typical animal-learning experiment. Put rats or pigeons in a *Skinner box*—a cube with a few lights, food dish, buttons, and wire relays—and set them going on a schedule of reinforcement where every press or peck on a button yields one food pellet. If the experiment is set up correctly, rats and pigeons can easily learn this task. Now change the reinforcement schedule: for every three presses, one food pellet is delivered. They also learn this, and more, including the rule "To get one pellet, press twenty-four times, no more, no less." If that doesn't impress you, then what about this: Give a pigeon three buttons. The left button provides food if pecked forty-five times, the right if pressed fifty times. When the center button lights up, the pigeon pecks away until the experimenter turns the light off. Next, both side buttons are illuminated. The pigeon must recall how many times he pecked before the experimenter turned the light off, and then peck the side button associated with this number. Thus, if the pigeon pecked forty-five times on the center button, he must turn and peck the left button. Pigeons solve this problem.

There's more. Teach rats to press one button for two light flashes and another button for four light flashes. No problem. Now, present either two or four sound beeps in succession. The rats immediately transfer their knowledge from the visual to the auditory modality, and do so even when the duration of the flashes or tones varies. All that seems to matter is the *number* of cues. In this study and many others, researchers have systematically varied the length of time a cue is presented, how hard it is to depress the lever, the amount of elapsed time between presentations, and the subject's hunger level. All these factors could serve as cues for figuring out when to press, and how often. Nonetheless, rats and pigeons ignore these cues, using the number of presses to maximize the amount of food obtained. They clearly have a number category. But do they have a number concept?[11]

Rats and pigeons are trained in the highly controlled environment of a Skinner box. By maintaining such conditions, the laboratory psychologist profits from exceptional control of factors that might contaminate performance and, thus, the interpretation of a subject's cognitive ability. In contrast, other studies carried out in the United States since the late 1960s have generally been guided by a different approach. Although extensive training was involved, animals such as chimpanzees, gorillas, dolphins, sea lions, and parrots were brought into a highly social environment, one involving direct human contact.[12] Most of this work focused on problems of language acquisition. Some studies, however, also explored conceptual domains outside language, while taking advantage of the quasi-linguistic skills that the animals had acquired. Experiments on the conceptual domain of number represent one of the more fruitful research ventures.

Alex is an African gray parrot. Irene Pepperberg, an ethologist, is Alex's primary trainer. Alex speaks English, sort of. He has learned, through several hours of training each day, for over twenty years now, to use English words to refer to such complicated concepts as *same*, *none*, *colors*, and *numbers* up to six. We know that Alex understands some of the principles of counting because he can give the correct answer to questions such as "How many X?" where X includes a cork, a blue plastic triangle, a piece of corn, and a toy truck. For example, when presented with five corks and asked, "How many?" Alex answers, "Five." This suggests that he understands the one-to-one mapping of symbol to object, property indifference, and cardinality (i.e., the last term labels the number in the set). But he understands much more.

In one experiment, Alex was presented with a tray consisting of several different objects such as four corks and five keys. He was then asked about the number of corks. Alex's answer: "Four." This shows that he readily transfers his understanding of number from the same class of objects to a mixed class. Neither Alex nor Irene stopped here. The next test was even harder, because it required Alex to attend to several conceptual distinctions at the same time. Alex sees a tray with one gray cork, two white corks, three gray keys, and four white keys. He is asked, "How many white corks?" He answers, "Two." Here, as in trials with other objects and properties, Alex must attend to both color and kind terms, and use both pieces of information to determine the correct number of items. He does so with apparent ease. Therefore, subitization—the counting mechanism that operates in the absence of focused attention—cannot account for Alex's performance. Alex is doing something more sophisticated, but it is not yet clear what.

Ai is an adult female chimpanzee. Tetsuro Matsuzawa, a comparative psychologist, has worked closely with Ai for over twenty years. Like Alex, Ai has also learned symbolic representations of number, but rather than vocal signs, she has learned the Arabic numerals. And, far exceeding Alex's talents, Ai knows the count sequence up to ten, including the Arabic symbol for zero. When presented with a string of three different numbers on a monitor, Ai points out the correct sequence, from lowest to highest, independently of the intervals between numbers (e.g., 1-2-9, 0-4-5). This shows that Ai understands the relationships between numbers, that a given integer does more than label a particular number of items. Each integer holds a particular relationship to other numbers in the count sequence. By carefully ruling out several alternative explanations for Ai's performance, Matsuzawa has demonstrated that at least one chimpanzee can acquire some aspects of the number concept.[13]

When we look at the performance of Alex and Ai, it is hard not to be impressed by their apparent abilities with number. However, if we look at the training needed to acquire such numerical skills, we learn something even more interesting. Recall that it takes children three to four years to learn the count sequence properly. However, once the child has learned the sequence one, two, three or one, two, three, four, she needs no more teaching or experience. She has acquired an understanding of a count sequence, a list of numbers that she can construct by adding one to the previous number. Each step in the training procedure was equally difficult for Ai and for Alex. For example, Ai needed about as long to learn the symbolic representation of 3 as she needed for 4, and as long for 4 as for 5, and so on. Thus, although Ai's performance on tasks involving discrimination of the numbers 0 to 10 is no different from that of a child of three to four years, the process of acquisition tells us that these two species represent number in different ways.

MATHEMATICAL ORIGINS

The evidence provided by Alex and Ai builds a strong case for the presence of the number category in animals. And knowledge of ordinality provides some evidence for a number concept. A series of experiments in the 1980s by the psychologist Sarah Boysen suggests that the chimpanzees' conceptual representation of number may be even more abstract than anticipated. Boysen spent three years training her star chimpanzee Sheba to learn the association between printed Arabic numerals (1 to 4) and the quantities to which they refer. Because the training incorporated many different objects, we can be confident that Sheba's understanding of number is abstract, and not tied to a particular item or class of items. The first experiment involved a test of Sheba's capacity to understand addition. Boysen hid some oranges in one box, and some more oranges in a second box. Sheba was trained to visit each box, and then report the total number of oranges discovered. For example, at the first location she finds two oranges, and at the second location she finds one orange. To solve the problem, she must calculate the total number of oranges and pick one of four cards labeled one, two, three, and four. Sheba performed well, picking the correct sum in a significant proportion of trials. The same experiment was then run again, but this time the cards with Arabic numerals replaced the oranges. Again, she was successful. When Sheba found a card with the number 2 on it, followed by a card with the number 1, she consistently picked the number 3 card at the final station.

Boysen suggests that Sheba can do addition by manipulating integer-like symbols. Although this is one possible interpretation, there are others. Sheba sees two oranges or the number 2 and opens up two files or slots in her memory banks. She then finds one orange or the number 1 and opens up another file. Now three files have been opened, and this matches the card with an Arabic number 3—the number of files that would be opened if she found three oranges or simply the number 3 card. In this case, there are no symbolic integers floating in her head. Rather, like the performance of human infants tested with Mickey Mouse dolls, Sheba's performance may derive from a relatively simple memory storage system. When objects are first seen and then hidden from view, a representation of each object is stored in a memory file. When all or some of the objects are presented, the filing system enables the individual to determine whether there is a match or a mismatch. This system works well with small numbers of objects, something on the order of three to four. To date, Sheba's numerical ability falls within this range.

The experiments with captive chimpanzees and at least one African gray parrot show that they can learn some aspects of a symbolic number system after years of training. What is less clear is how such symbols are represented, implemented, and used in tasks involving numerical computation. In the absence of this information, we still can't claim that animals have an understanding of the number concept, at least as we understand it for humans. Before we delve into this question, however, let's take a step sideways and explore whether animals might use their numerical abilities during social interactions.

An understanding of number as either category or concept would certainly come in handy if one were negotiating over access to food. For example, when chimpanzees share meat after a successful hunt, do they keep track of the number of pieces of meat dispensed? Do they punish those who fail to share? Boysen developed an elegant experiment to see whether chimpanzees might use their understanding of number during food negotiations. Here is the game, played between two of the world's best-trained chimpanzees: Sarah, who was tested in hundreds of experiments by the psychologist David Premack, and Sheba, Boysen's star. One individual plays the role of selector and the other, the role of receiver. Let's say that Sheba is the selector and Sarah the receiver. Sheba is presented with a tray containing two containers filled with food treats; one container always has more treats than the other. Sheba points to one container. The amount of food in this container goes to Sarah, and Sheba is left with the food in the other container. If chimpanzees are greedy, and we assume they are, then Sheba should always point to the container with fewer treats. In so doing, she guarantees receiving the container with *more* treats. What happens? Sheba always points to the container with more treats. If she is presented with one versus six, she picks six. If she is presented with two versus five, she picks five. Sarah always wins the larger quantity! And the same thing happens when Sarah plays selector and Sheba plays receiver. Either they just don't understand the game, or our assumption about greed is wrong. Indeed, perhaps chimpanzees are wonderfully altruistic, Gandhis of the animal world!

The key test comes next. Sheba, you will recall, knows her Arabic numerals. Instead of presenting Sheba with a tray of treats, researchers present a tray where each well is covered by a card representing a distinctive Arabic numeral; for example, one well is covered by a card with the number 1 and the other well by a card with the number 6. Under these conditions, Sheba picks the 1 card. Consequently, she receives six treats while Sarah receives one treat. Beautiful. This result shows that when the temptation of food is removed and replaced by the equivalent number card, Sheba passes with flying colors. In terms of her understanding of number, it shows that she understands the property of ordinality, that any given number is *greater than* or *less than* some other number. But this result also tells us something else. Chimpanzees, at least these two well-trained ones, have difficulties solving a task that requires them to inhibit one response in favor of another, more favorable response. I will say little else about inhibition here, but encourage readers to keep it squarely in the frontal part of their brain—this is, in fact, precisely where the problem of inhibition starts and where we retrieve information over the short run.[14]

We know from research on number in infants and children that a substantial proportion of their knowledge emerges spontaneously. But all the results on number in animals that I have reviewed depend on substantial training. We must therefore ask whether animals spontaneously represent number as a category or concept.

Working with some of my students, I set up a magic show for the rhesus monkeys living on the Puerto Rican island of Cayo Santiago. We started with a virtually identical version of Wynn's $1 + 1 = 2$ task for human infants. Rather than Mickey Mouse dolls, however, we used bright purple eggplants. After familiarizing the rhesus to the eggplants and display box, we presented them with the test trials. In each trial, subjects watched as an experimenter placed two eggplants behind a screen and then removed the screen. Subjects looked longer when the test outcome was one or three eggplants than when it was the expected two. Like human infants, rhesus monkeys appear to understand that $1 + 1 = 2$. Rhesus monkeys also appear to understand that $2 + 1 = 3$, $2 - 1 = 1$, and $3 - 1 = 2$. They fail, however, to understand that $2 + 2 = 4$. Comparable results have been obtained with adult cotton-top tamarins.

The monkey experiments indicate that "looking time" is a useful measure for at least two nonhuman primate species, and will probably work with many more species. Because of the diversity of species that could be tested—all they need are eyes and interest in visual displays—the expectancy-violation procedure provides a powerful technique for studying what animals know— a technique that is so simple that you could try it out at home with your pets. This technique differs from many others in that it reveals what animals think, spontaneously, in the absence of training. With respect to numerical abilities, adult rhesus monkeys and tamarins are at least as talented as one-year-old human infants when it comes to summing objects. Spontaneous representation of number in these primate species is limited to a small number of objects. With training, animals surpass this limitation to some extent. Human children also exceed this early limitation, and ultimately leave animals behind in their numerical sophistication. At present, we do not know which mental tools fuel this developmental change in human children. Several researchers consider language acquisition to be essential, providing children with a mechanism for establishing a count list and organizing the conceptual properties of the number system. Others consider language a facilitator of the child's knowledge of number.[15]

NUMBER CRUNCHERS

What kinds of evolutionary pressures might have favored greater numerical competence, more sophisticated discriminations and computations?[16] I suggest that in nature, animals confront situations where relative rather than absolute quantification is required. In amphibians and fish, where a large number of eggs are laid and guarded, parents lack an understanding of the precise number of eggs laid, the number that have died, and the number that have been removed by a predator. Many avian species have their nests parasitized by members of another species such as the cuckoo. Rather than raise its own young, the cuckoo deposits its eggs in a host's nest, allowing the host parent to incubate and feed the young. Sometimes the parasite does a one-for-one swap, knocking out the host egg and replacing it with one of its own. In other cases the parasite adds eggs without removing a host's egg, or the number added exceeds the number removed. Although there are number differences, there is no evidence that hosts shift their allocations of parental care. They don't seem to be counting at all. For most species, then, parental investment appears to be guided by approximations rather than specific numbers.[17]

When animals compete for resources, the number of individuals they track is small: a few competitors of higher or lower rank, two or three allies in a coalition, and a small number of potential mates. In a troop of fifty baboons, individuals might notice the disappearance of a troop member, but are unlikely to think, "Geez, we're down to forty-nine. That puts us at a disadvantage against our neighbors who have fifty."[18] It seems highly unlikely, therefore, that animals living under natural conditions would confront ecological problems that would select for greater numerical competence. This claim does not, of course, rule out the possibility of showing more exceptional numerical abilities through training in a laboratory environment.

If animals are limited in their ability to carry out numerical wizardry, what mechanisms evolved in our most immediate ancestors that enabled them to step to the head of the class with respect to this particular mental tool? The answer, highly speculative at this point, is tied to a problem that we shall discuss in chapter 8. Here's a preview.

Humans are endowed with two cognitive talents that animals lack, at least naturally: first, the capacity to assign, spontaneously, arbitrary symbols to objects and events in the world; and second, the ability to manipulate the sequence and order of a string of symbols to alter their meaning—a combinatorial system. The explosion in numerical competence is due to the child's capacity to formally manipulate symbols. Along with the capacity to take a growing lexicon and manipulate words, the child acquires the ability to juggle number symbols. Although some of the basic elements of the number system (e.g., property indifference, individuation of objects) are in place before the basic elements of language, more abstract elements of number (count sequence, symbol mapping) emerge after the child has acquired a reasonable command of words. In this sense, a child's numerical abilities are immature when compared to her linguistic abilities. The pattern of development is like a game of leapfrog, with some aspects of our numerical competence emerging before our linguistic competence, and some aspects emerging afterwards. At present, we do not understand how these two domains of knowledge affect each other, either during the course of evolution or during development.

The combinatorial engine underlying our number and language systems allows for a finite number of elements to be recombined into an infinite variety of expressions. However, the evolutionary origin of this capacity remains unclear. Did it evolve for number, language, or both? Clearly, the number system of animals shows no sign of combinatorial power. The natural communication systems of animals show no sign of combinatorial organization either. At present, therefore, research on animals does not help our understanding of this evolutionary problem. Developmental data on children, however, help a bit. Given that children's grasp of counting does not develop until well after they produce sentences, it would appear that recombination is first tapped by the language system and then, somewhat later, by the number system (see chapter 8). Studies of brain-damaged patients show that some individuals have linguistic deficits but no deficits in numerical competence. Conversely, some individuals have numerical deficits but no linguistic problems. This suggests that separate computational systems are responsible for language and number.

What I propose, therefore, is that the selective pressure responsible for the emergence of a combinatorial system, one that allowed ancestral humans to enumerate at a more precise level than any animal, is the emergence of exchange systems—trading, to be precise. Whether you are trading spears, mongongo nuts, goats for a dowry, or coins, you want to be sure that you know how much you are getting, and that it is a fair exchange. Approximations are doomed to failure in this kind of system. Though some animals engage in reciprocal exchanges (see chapter 9), they are not based on any kind of quantitative precision. Vampire bats regurgitate blood to those who have regurgitated to them in the past, but they aren't counting the number of milliliters. Bonobos trade food for sex, but they don't count the amount of food dispensed nor tally the number of copulations obtained. In all these interactions, the system works on the basis of approximate returns. When social exchange of material goods emerged onto the scene, selection favored those individuals capable of enumeration, carrying out combinatorial computations with symbols. Why early humans demanded precise reciprocal exchange, whereas animals tolerated approximate returns, is a mystery. One possible answer lies in the evolutionary history of our moral sense (see chapter 9) and the intuition that humans, but no other animal, place values on objects and actions. This leads to a sense of fairness, equipped with the ethical judgments of right and wrong.

The third, and final, installment in the universal toolkit is
a mental tool that allows all animals to navigate. All animals have this basic tool, with specializations evolving in
the face of unique environmental problems.

4

Space Travelers

The Tunisian desert ant, after finding food, travels hundreds of meters straight back to a dime-sized nest entrance, traversing sand dunes lacking road signs for home. On a yearly basis, the garden warbler revisits the same breeding area in central Europe and then migrates to tropical Africa to revisit the same overwintering site. Chimpanzees leave stone hammers in areas of dense rain forest vegetation where they will be of use when the palm nuts ripen. Clark's nutcracker, a food-storing bird, can hide up to 33,000 seeds in over 6,600 locations in the fall and then return months later to retrieve most of them. Salmon travel around for several years in the Pacific Ocean, covering thousands of kilometers, and then return with spectacular accuracy to the streams in which they were spawned. Alongside these cases of navigation by wild animals are hundreds of observations of pet dogs and cats accomplishing equally spectacular journeys to find food or their beloved owners.[1]

These cases illustrate that animals are equipped with a mental tool for spatial analysis, each tool outfitted with customized features for solving the particular demands of their environment. To understand both the similarities and differences among animals, we need to determine how individuals represent the relevant space before a course has been charted, while a path is being traveled, and after the destination has been reached. Animals don't have computer-generated maps at their disposal, but they do have other tricks. When the desert ant has drifted five hundred meters from home and looks out at the barren topography of the surrounding dunes, it determines the shortest path home by both referring to a map of the sky that is stored in its brain and making adjustments in the direction and distance traveled. The ant works like the great explorers of the past, such as Columbus during his exploration of the New World, and Marco Polo during his journey to Asia. In the absence of sophisticated maps, they too used the celestial patterns and their keen sense of direction and distance to navigate through uncharted waters.

When a Clark's nutcracker retrieves the goods from its widely dispersed stash, does it fixate on crucial landmarks such as trees, bushes, or bends in a river? If landmarks are used, how are they represented in memory? A bird might stash a seed next to a tree at the beginning of winter, but that same tree will look different in the spring. This suggests that for a landmark to be useful, its featural details should not be stored in memory, especially since nature has a way of changing its morphology. The animal must be capable of identifying and reidentifying the object or location even when its appearance has been transformed. In this sense, the animal's representation of a landmark depends on some of the mental tools that underlie knowledge of objects and number. For example, if one is to identify something as a hammer or as a member of the category "seven," some featural changes are irrelevant with respect to classification (e.g., the *color* of a hammer or the *kind of object* enumerated), whereas other features are highly relevant (e.g., a *shape* change that turns the hammer into something that looks like a screwdriver, or a change in the *number of objects* from seven to six). Similarly, whether the tree next to your nest has leaves, is barren, is covered in snow, or is filled with a flock of birds, it remains the same tree and, thus, the same useful landmark. Animals, including the nutcracker, appear to make use of this fundamental property of landmarks, recognizing that identity is preserved across some featural transformations.

To find their way home, locate food, or find a receptive mate, some animals depend on their memories more than others. As a result of these imperatives, structures in the brain have become specialized to facilitate such memory-dependent tasks. In promiscuous rodents, where males range far and wide to find sexually receptive females, the hippocampus—a part of the brain that helps handle the memory part of the navigational problem—is much larger in males than in females, and males perform better on tests that tap their spatial powers. In monogamous species that are otherwise closely related, there are no sex differences in spatial ability or hippocampus size. In food-storing birds, the hippocampus is larger than in closely related birds that do not store food. In some parasitic cowbird species, females carry the burden of finding host nests for dumping their eggs, and have a larger hippocampus than do males. And some researchers have argued that sex differences in human spatial ability represent a reflection of the ecological pressures that our ancestors experienced, the difference between hunting males and gathering females, a legacy of sexually promiscuous males roaming the savanna in search of females.[2]

This chapter examines the last mental tool in the universal toolkit. By exploring the diversity of navigational systems that have evolved, and assessing the causes of convergence and divergence among species, we will extract the crucial principles that guide the earth's space travelers.

SOLO EXPEDITIONS

In *The English Patient* and *Smilla's Sense of Snow*, two adventure novels transformed into big-screen extravaganzas, the leading characters are presented with significant navigational problems: they must locate someone in danger by traveling across featureless terrain, sand dunes and snow fields, respectively. What makes such spatial navigation difficult is the lack of significant landmarks. As we have already learned, however, the desert ant routinely cracks this problem with apparent ease. How?

For the desert ant, there are no designated feeding areas, no food oases. When an individual leaves its nest, therefore, it travels in a convoluted route in search of something to eat. For an animal less than a centimeter in length, its home range is vast, extending out from the nest some four hundred or so meters. Unlike Hansel and Gretel, desert ants do not leave trails behind. Although they certainly leave traces in the sand, in terms of both a depression and a scent, neither last long enough to be useful. More important, retracing the path to food is unlikely to represent the shortest path back to the nest. Given that animals have been designed to maximize energetic returns from food intake and minimize travel costs during search, finding the shortest path during navigation is highly adaptive. For the desert ant in particular, minimizing the distance traveled is of utmost importance, since they can survive outside the nest for only about two and a half hours in the hot summer sun. Detailed studies over the past twenty years by the biologist

The Tunisian desert ant's circuitous travel route out of the nest to find food, followed by ant-line back to the nest by means of dead reckoning.

Rudiger Wehner reveal that these ants take the shortest route home once they have located the target object. The same can be said for honeybees, goslings, gerbils, turtles, dogs, and humans.[3]

There are at least two recognized mechanisms for calculating the shortest path between two points that are out of sight from each other. I say "at least two" because there might be different mechanisms operating in species that sense the world through modalities that we are unfamiliar with, such as magnetic and electrical. One recognized mechanism involves forming a *cognitive map* of the home range area. This map, used to plan and implement a journey, represents a record in the brain of how key points in the environment are geometrically related. What is crucial about this representation is that it is highly flexible. A map-carrying individual can readily find the shortest path from A to B inside the targeted area, even if the route traveled is novel. Imagine that your map consists of a baseball diamond. You have traveled the path from each base according to the rules of the game, home plate to first, first to second, second to third, and third to home. During a practice session, you are on third. Your coach tells you to close your eyes and walk to first base. Although this is a route you have never traveled, and quite a bizarre request, it represents a trivial task. In your mind, you know that first base is to your left, on a straight line through the pitcher's mound. You don't need to see where you are going and you certainly don't need to go back to second and then over to first. The geometry of the situation sets up the shortest path, and the cognitive map in your head allows you to tap this geometry.

A second way to calculate the shortest path between two points is what both seafaring navigators and scientists refer to as *dead reckoning* or *path integration*. Using this system, individuals continuously update the angle of change from their departure point, as well as the speed and distance traveled. As in any sampling procedure, however, there are trade-offs between the number of times one samples and the speed with which one is able to reach the targeted destination; nonstop sampling will certainty reduce the probability of making an error, but will also increase the amount of time required to reach home. What is the evidence for these different mechanisms?[4]

When the desert ant leaves its nest, it calculates its direction by the sun's position in the sky. The ant can make use of the sun's position because it has a reference map of the celestial patterns in its brain. This map, part of the ant's neural equipment, allows it to learn about the relevant changes that occur as a function of where an individual lives and forages. Thus, as the ant travels away from its nest, it enriches the celestial map in its brain by acquiring information about the angle and distance of displacement. This acquisition process requires active movement rather than passive receipt of spoon-fed information. What would you imagine happens to an ant if it is picked up following emergence from the nest and deposited several meters away? Given its presumed familiarity with the local turf, you might expect it to be perfectly clear about its whereabouts. In reality, however, the ant travels around aimlessly, rarely finding the path home. Ants require personal experience with the travel route in order to register the relevant information about displacement, speed, distance, and time. Their celestial map is insufficient, as are the "You are home" landmarks immediately surrounding the nest. These landmarks are, however, of use when the ants shift from a dead reckoning mode of navigation to *piloting*. Piloting makes use of local landmarks and is employed by the ant (as well as other animals) as it approaches within a few meters of the nest entrance, a highly familiar area.

Animals that use the sun or any other celestial body to navigate must recognize that the pattern perceived is not static. They must represent changes in the position of the sun or another star over time, and calibrate such knowledge with their own sense of time. Because the sun is constantly moving, its relative position in the sky provides a potential cue to the time of day.

To appreciate the impact of the sun on animal navigation, try the following experiment next time you see an ant on a sunny day. As the ant is traveling along a path, use a mirror to reflect the sun's image to the opposite side of the ant's body. You will most likely observe an about-face, a navigational reorientation that allows the ant to maintain its relative angle to the sun's azimuth. You can play this game all day. The ant won't catch on. Like many other animals, the ant relies on the sun's position while navigating with dead reckoning. In fact, the sun's virtual control over navigation is so complete that even in the face of landmarks indicating that the ant is moving in the wrong direction, it continues to rely on the information provided by the sun's position.[5]

To understand how spatial navigation is influenced by the sun's position and the animal's ability to time events, the ethologist Jim Gould transferred resident beehives from one location to a new area within the same time zone; this manipulation eliminated their ability to use local landmarks to travel home, but preserved the animal's sense of time. One group of bees was kept in the dark one hour before noon, whereas a second group was trapped at noon and kept in the dark for one hour. On their first trip out of the hive, the travel route of the pre-noon group suggested that they had underestimated the sun's azimuth. The noon group had overestimated it. The bees behaved as if the sun's rate of movement at the time they were trapped had continued during their confinement. In fact, the relative speed of movement of the sun had increased for the pre-noon group and decreased for the noon group. This pattern makes sense if bees track the relative rate of change in the sun's azimuth during the day and calculate a running average of its position over a particular period of time. This is an effective mechanism because the sun's position, like many other physical properties, is a statistical regularity. When such regularities arise, natural selection favors innate circuitry that rapidly processes the relevant information.

To determine how an animal's timing system might influence its ability to navigate, we need the experimental equivalent of a cross-continental flight, one involving a time zone shift and jet lag. In one experiment, bees trained to fly from their hive to a feeding station hundreds of meters to the northwest were transferred overnight in their hive to a time zone three hours behind. When released from the hive, the bees searched too far west, as if their internal clock was telling them that it was three hours later. The information gleaned from the sun's azimuth conflicted with the information already programmed in the bee's brain.[6] What this tells us is that spatial navigation is a multimedia adventure, involving simultaneous integration of information from several sources. How, then, do animals rank the different sources of information during their travels?

When we examine the relative contribution of different kinds of information to spatial navigation, we find that birds are an interesting test group. For some birds, the use of celestial patterns (sun during the day, other stars at night) is likely to be far more important than for nonmigrant birds that remain within a relatively confined territory throughout the year. Such species include the garden warbler, which crosses the equator while migrating over continents, and the homing pigeon, which travels over long distances without familiar landmarks.[7] All birds, however, share a common developmental pattern and a common problem: they start off life in a nest, with time on their hands to stare up at the sky and collate information about where they are relative to the celestial patterns above. But the knowledge they acquire from this early exposure to the skies is of no use until much later when they fledge and move out on their own. The problem for young birds, then, is that they process information about the skies early in life, store it for up to a year, and then access the data when they take their first flight. What celestial information must the young bird acquire?

Consider the indigo bunting, a bird that lives in the temperate zones, is born in the spring, and then migrates to warmer areas during the fall. In the 1970s the biologist Steve Emlen raised buntings in a planetarium, providing groups of birds with access to different constellations. The constellations on the planetarium walls rotated during the period of learning. As long as the buntings could see a patch of sky on the planetarium walls during the learning period, they were able to orient correctly on the autumnal migration. They misoriented, however, if they were given no experience with a constellation (reared in a skyless world), or if the particular constellation they were exposed to was blocked at the start of their migration.

Emlen's work on buntings shows that the stellar map is constructed before the young bird leaves the nest, but only if the relevant experience is processed during a specific window of opportunity, what biologists call a *critical* or *sensitive period*. As such, the representation of celestial patterns is like the representation of song in birds and language in humans (see chapters 6 and 8). Each of these systems is guided by a set of instincts, innate mechanisms that cause the developing organism to attend to the right things at the right time. Importantly, however, the way a bunting acquires song and the way it acquires spatial knowledge are different, involving different brain structures and developmental timetables. This contrast reinforces a point I made earlier, that for many domains of knowledge, natural selection has favored highly specialized learning tools. If the organism depended on a domain-general learning system, one that was blind to the kind of experience or knowledge to be acquired, it would often learn the wrong things, or learn the right things too slowly, and thus decrease its chances of surviving.

The bunting results also raise an interesting point about the nature of spatial representations. Each night presents a slightly different picture of the sky due to variation in cloud cover, season, illumination from the moon, and vegetation surrounding the nest. Consequently, animals must be capable of tolerating some of the variation they see, while preserving their ability to recognize stellar configurations. Each night, when the young nestling looks up to the sky, it effectively gains new information about celestial configurations. Sometimes it is facing north, and there is extensive cloud cover over the Big Dipper, whereas sometimes it is facing south, the sky is clear, and every bit of the Big Dipper is visible. If the nestling depended on a fixed mental representation of the Big Dipper—a total of seven stars with the center star sitting directly above Polaris at 9:00 P.M.—it would often make mistakes when using this constellation to navigate home. To be useful in navigation, therefore, mental representations of celestial bodies must be flexible, allowing for variation in what can be seen when and where.

Many animals travel by referencing the sky. Depending on the scale of the environment, they also use landmarks during their travels, sometimes to find home, and sometimes to figure out the proper departure path. How are such landmarks represented, stored, and accessed during spatial navigation?

Among the many demonstrations of landmark use, one of the more elegant experiments was conducted in 1997 using Clark's nutcrackers. This is an excellent study animal for a test of spatial knowledge because its survival depends on spatial analysis in its work of storing and retrieving massive numbers of seeds. The psychologists Alan Kamil and Juli Jones trained Clark's nutcrackers to find food at a midway point between two landmarks, positioned on a north-south axis and separated by distances of 20 to 120 centimeters across trials. Then the researchers varied the distance between landmarks, while continuing to place food at the midway point. The nutcrackers readily solved this task, searching at the midway point regardless of the distances. Furthermore, the accuracy of their searches was only mildly influenced by a rotation of the landmarks away from the north-south axis. These data show that nutcrackers form a representation of the geometric relationship among landmarks—something like *the middle*—and use this to find stored food.[8]

This finding is important because it shows that animals can tap into fairly abstract concepts about space when navigating, using landmarks to anchor the spatial coordinates. Unlike navigating by means of specific landmarks which provide direct, perceptually salient details—a rock, tree, or flower patch—referencing the *middle* is abstract, relatively detached from the specific details of the environment. If an animal has learned that food is in the *middle*, then it should be able to find food in the middle of any two objects, any distance apart, and along any directional axis. If an animal such as the Clark's nutcracker can search by reference to the *middle*, then perhaps it or some other species can search by reference to *above, below, to the right,* and *inside*. Such tests have yet to be run.

The nutcracker experiments are important on another level. Some cognitive scientists working on human navigation have argued that our own sophistication as a species derives from language, from having words for the concepts of *above, below, to the right*, and *inside*. Support for this claim (see the section "Intelligent Errors," this chapter) comes, in part, from studies of the developing child. For example, studies conducted in the 1990s by the developmental psychologists Linda Hermer and Elizabeth Spelke show that children's spatial knowledge is impoverished relative to that of adults. This difference apparently disappears once the child acquires a command of words for spatial concepts. Like many hypotheses that see language as necessary for certain kinds of thought, this one is also based on a correlation: spatial knowledge is acquired at the same time as linguistic knowledge. But correlation is not causation. We cannot infer that spatial words improve spatial navigation. In fact, language may play no direct role at all. More important, given that the linguistically inept nutcracker can navigate by referencing the abstract concept of *middle*, it appears that language is not necessary for such spatial knowledge at all. This argument leaves open, of course, the possibility that in humans, language enhances our spatial knowledge. But this is a much less controversial claim, and one that most researchers working on animals would be happy with.

Animals clearly rely on visual information to navigate. But are other modalities functionally significant and necessary? If visual cues are removed altogether, can input from the other senses compensate by building up a sufficiently rich representation of the spatial surroundings that the individual can maintain an accurate navigational course? Experiments on pigeons and salmon guided by smell and a magnetic sense, as well as work on blind rats and blind children, provide the answers.

Studies of homing pigeons indicate that they use both visual and olfactory cues to find their way. When researchers use a local anesthetic to temporarily eliminate the capacity to smell, pigeons initially orient in the wrong direction when released from an unfamiliar area, but have no difficulty fixing their course when released from a familiar area. Further, when heavy-duty fans are used to alter the smells in the air, pigeons experience navigational difficulties, often missing their loft on a return trip. This suggests that pigeons have an olfactory map of sorts, and use this during navigation.

Salmon also use smell to navigate, and are perhaps even more dependent on odors to find the target location. Early in development, young salmon imprint on the smell of their home stream in much the same way that young chicks imprint on and then follow the first moving object that crosses their path. After this juvenile phase, however, the salmon move out into oceanic waters, thousands of kilometers from home. When it is time to spawn, they return to their birth site, using a variety of clues to find the way. Like other species, including rainbow trout, salmon can use the sun's position to navigate in the ocean. Because of magnetically structured crystals in their brain and the lateral line of their body, however, they can also navigate by detecting changes in the earth's magnetic fields. When salmon approach the spawning streams, however, they are guided in by highly familiar olfactory currents.

One might expect that in mammals vision would be a dominant, perhaps even crucially necessary, modality for navigation. Experiments on maze running in rodents show, however, that blind subjects do as well as sighted subjects, tapping into their backup tactile system for feedback. Among blind children, use of words that mark spatial locations (e.g., *right, left, in, on, top, below*) appear in the vocabulary at about the same time as they do for sighted children. To determine whether humans, like rats, are also capable of compensating for the loss of visual information, the psychologists Barbara Landau, Elizabeth Spelke, and Henry Gleitman collected observations in 1984 from a congenitally blind child. At around the age of two years, the child was brought into a test room with her mother. The mother and child formed the point of a diamond-shaped configuration, with a basket to the right (first base), a table straight ahead (second base), and a pillow to the left (third base). The child was first trained to walk back and forth from Mom to the basket, from Mom to the table, and from Mom to the pillow. The child was then brought to one of the bases and asked to use the shortest path to a different base; this task requires the child to imagine novel routes, such as walking from the pillow to the basket. The child's travel routes indicated that she had succeeded, taking the shortest path to the target on a majority of runs.

Animals, including humans, can maintain a course of travel using information from sensory modalities other than sight. Although visual cues may dominate for most mammals, their navigational system depends on the contributions from multiple sensory modalities, each providing information to orientation, path finding, and target location.[9]

MAPS IN THE MIND

Maps, like ancient scripts, represent records of minds from the past. In 1870 a collector specializing in rare trinkets was walking down a street in Modena, Italy, and happened to notice a map hanging on a wall in a grocery store. But it was no ordinary map. Rather, it was a rare nautical map created during the Renaissance, one that charted new routes from the New to the Old World, including some of Columbus's explorations. The man convinced the store owner to sell the map. With prize in hand, he headed to the library and made a donation. To this day, the famous Cantino Map hangs on the wall of the Modena library, providing a window into how humans thought about the world, how they represented in their mind a map of the world.

The integration of multiple sources of information that we discussed in the preceding section brings us back to the problem of cognitive maps. Thus far we have shown that a diversity of animals navigate by means of dead reckoning. This is a universal tool, one that appears in the mental toolboxes of all animals. Controversy arises, however, in considering the possibility that animals use a cognitive map to select optimum travel routes. A cognitive map represents an extra device in the spatial domain, an add-on to the universal spatial tool.

Cognitive maps are representations in the brain of the geometric relationships between salient sites in an animal's environment. The animal forms the representations by acquiring information about the coordinates or positions of landmarks and relevant scenes. Compared with dead reckoning, navigation by reference to a cognitive map provides considerable freedom with respect to finding optimum routes, even when the starting point and target are completely novel. If the map also includes information about distance and direction, it represents the shape of the environment, its perimeter and topography. For some scientists, the crucial test of whether an animal navigates by a cognitive map is its ability to take novel shortcuts. For others, evidence of cognitive maps requires using the remembered location of the target area relative to landmarks. We don't have to worry about which definition is better or more important. In fact, many researchers in this field would be happy to throw the term "cognitive map" out the window. The important point is that we must determine the kind of representation underlying animal navigation.[10]

The experimental work on the congenitally blind child provides one example. To maintain a straight course from one location to another, the child used dead reckoning, managing the distances and directions traveled. But to orient in the correct direction from the starting position, she had to make use of a cognitive map, a suite of geometric relationships between Mom, the pillow, the table, and the basket. Work on the Clark's nutcracker provides another example, indicating that these birds can find the midway point between two landmarks, an abstract geometric relationship.

Let's return to honeybees. Like the desert ant, honeybees depart from a home base in search of food. But for honeybees, finding food depends less on luck and more on sampling from relatively well known foraging sites, areas where food availability depends on seasonal variation in pollen. When a honeybee forager returns and dances, other hive mates pay attention. Depending on the information in the dance and the current needs of the hive with respect to finding food as opposed to storing it, the observers will either stay put or go out on their own foraging expedition. The observers must therefore process the information in the dance and then place it within a system of spatial representation.

The important question is, of course, what kind of spatial representation underlies the hive mates' decision to forage, and if they depart from the hive, how do they find the food? For some researchers, the fact that honeybees find the shortest route home when released from any site within a familiar home range is evidence that they are using a cognitive map to navigate. They have a system in their brain for manipulating symbols associated with landmark coordinates. For others, such observations are taken as evidence that bees navigate by recalling previously traveled routes; when they leave the hive to forage, they carry a mental Polaroid camera around with them, snapping off shots of highly traveled routes. Although controversy in this area continues, my own sense is that the weight of evidence tilts toward the cognitive map position for one simple reason. For the snapshot view to work, honeybees would have to store an extraordinary number of images of familiar routes. In the absence of such a database, they would be incapable of flying straight home from the variety of release sites that devious experimenters have set up. But I'm getting ahead of myself.

Gould observed a hive that had been maintained near a lake for a long period of time. This provided some insurance that the honeybees were familiar with the local environment. Each day, one group of foragers was trained to move from a release spot away from the hive to a boat on land, stashed with nectar; once they loaded up on a meal, they were captured and prevented from returning to the hive. Over the course of several days the boat was displaced further and further from the release site until one day it was square in the middle of the lake. At this point, the foragers were allowed to collect nectar from the boat and then return home. When the foragers arrived at the hive, they danced, indicating the location of the nectar-laden boat. Although the hive members paid attention to the dance, virtually no one flew out of the hive. Gould suggests that the honeybees responded to the foragers' dance by referencing their cognitive map. And for this colony, the map fails to reveal a "Food Here" sign in the middle of the lake. Skeptical of the dancers' message, hive members wait for a more reliable dancer.

Do these observations seal the argument in favor of cognitive maps? Not quite. Perhaps honeybees avoid traveling across large bodies of water due to unpredictable air currents and other navigational challenges. If so, then their failure to fly toward the boat can be accounted for by their general avoidance of water. If this interpretation is correct, then they should never cross water to get to a target location. Rather, they should take a more circuitous route.

To control for the possibility that bees avoid crossing water, Gould displaced the boat an equal distance from the original release point, but kept it on land, at the edge of the lake. Here, following the foragers' dance, a large number of observers flew off to the nectar-rich source, crossing the water in a direct shot to the boat. Together, these results show that honeybees process the relevant information in the dance and evaluate this information relative to what they have remembered about the coordinates of the lake and the land surrounding it. In this sense, the honeybees' travel route is consistent with the idea that they are navigating by reference to a cognitive map, one that lays out the geometry of salient points in the environment.[11] Is there comparable evidence for other species?

Our understanding of navigation in monkeys and apes is, in many ways, worse off than it is for insects, birds, and rats, especially when it comes to providing clear evidence of the representation underlying their navigational patterns. In the Tai Forest of the Ivory Coast, chimpanzees use stone hammers to crack open palm nuts (chapter 2). Because good hammers are hard to find, and heavy to carry around when palm nuts are unavailable, chimpanzees often leave their hammers in key areas and then return to retrieve them at a later date. The ethologists Christophe Boesch and Heidwige Boesch suggest that these observations provide evidence that chimpanzees form cognitive maps of their environment and use them to retrieve hammers. There are, however, alternative explanations. Specifically, given their familiarity with the forest, their ability to dead reckon, and their presumed knowledge of salient landmarks (especially fruiting trees associated with a stashed hammer), they can combine dead reckoning to stay on course with landmark-to-landmark movements to find the hammer. This doesn't quite prove that chimpanzees represent the geometry of key points in the environment. Moreover, there is no evidence that they take the shortest path to the hammers when they sense that the palm nut season has arrived.

In experiments conducted by the psychologists Randy Gallistel and Audrey Cramer with vervet monkeys, and by Emil Menzel with chimpanzees, subjects were taken on a tour of an enclosure, and along the way, the experimenter indicated the location of a dozen or so feeding sites. When released, subjects followed a route that was nothing like the one on which they were initially brought. Both species navigated along a route that minimized travel distance. These observations suggest that individuals planned a travel route and did so by recalling the relative geometry of the baited sites. One problem with this interpretation, however, is that what appears to be route planning may actually be a series of sequential decisions. For example, the individual might arrive at the first baited site and after eating there, figure out its next move by referencing the sun's position and the nearest landmark. Here, there is no planning. No map. The animal is simply jumping from one point to the next. Without additional details of the environment in which these animals were tested, we cannot work out the precise nature of their spatial representation.[12]

No animal has yet created a representational map *outside* its brain, a physical drawing with clearly depicted routes and landmarks. In the 1970s, however, the psychologist David Premack considered the possibility that a chimpanzee might be able to understand and use a map created by humans. In the first experiment, a chimpanzee watched as an experimenter concealed a food treat under a doll-sized piece of furniture located in a three-dimensional miniaturization (model) of a full-sized room. The experimenter then led the chimpanzee to the full-scale replica of the miniaturized room and allowed the chimpanzee to search for the food. If the chimpanzee realized the correspondence between these two rooms, then it should have no problem finding the food. All the chimpanzees failed this test. By the age of approximately four years, however, human children easily solve this scaling problem, going from the miniature room to the full-scale one, as well as the other way around.

Not quite willing to give up, Premack ran another series of experiments. First, a chimpanzee watched as an experimenter baited one location inside a full-scale room. Next, the experimenter brought the chimpanzee into an identical full-scale room and allowed it to search. Under these conditions, the chimpanzees found the food. Gradually, the replica room was reduced to a model on a piece of canvas, with the ultimate step being a reduction to a map of the room, including all the key landmarks. The chimpanzees failed this task as well, showing that they do not understand the representational properties of the map, that there is a perfect one-to-one correspondence between landmarks in the room and landmarks on the map.[13]

There is no convincing evidence that animals form and then access the spatial representation of a shortcut. Animals do, however, store and then remember the geometric coordinates of relevant landmarks and surface features of the terrain, and use these representations to navigate. In this sense, animals form cognitive maps and use them in the service of navigating to target areas.[14]

INTELLIGENT ERRORS

When I first moved to Cambridge, Massachusetts, I walked and rode my bicycle everywhere. As a result, I eventually learned some critical landmarks—mostly restaurants, coffee shops, bookstores, and movie theaters—and generated a mental map of how to move efficiently from one destination to another, even when the path was novel. Given the snow and rain, I decided that my feet and bicycle were not always convenient, and thus purchased a car. What a disaster. My mental map of the area was useless. All the routes and shortcuts that I had been accustomed to were obsolete given that most of the streets in Cambridge are one-way. No problem for travel by foot or bicycle, but a serious impediment for drivers. Large sections of my map were deleted as I began the process of reconstructing new landmarks and new routes. It took another year for a car-friendly map of Cambridge to emerge in my head, and I am slowly adding on Boston and the neighboring areas. Fortunately, my wife and I are both spatially challenged, so we have come to expect and somewhat enjoy the process of getting lost.

Landmarks have played a significant role in many of the examples discussed in this chapter. To use a landmark for spatial navigation requires memory for what it looks like, where it is relative to other objects, and whether it moves or stays in one place. In this section I discuss how animal minds represent landmarks, and how a smart navigational system can look remarkably stupid. You will be amazed to learn, as I certainly was, that rats, pigeons, and human children literally trip over, bump into, and generally ignore landmarks under certain conditions. But such stupidity has a flip side. Ignoring a landmark is the result of a specialized mental tool, a domain-specific system of knowledge that will make intelligent errors because it is guided by a set of highly specific principles, theories, and hypotheses for how the world works and what to learn from it. But the cost of this kind of myopic learning system is that it will sometimes ignore the obvious.

Landmark-guided navigation requires the ability to learn about the appearance of relatively stable objects or visual scenes.[15] An object is useful as a landmark only if, when viewed from different perspectives or with certain features altered, it can nonetheless be recognized. There has to be some mechanism for pattern recognition, for matching up a stored representation with current reality. In honeybees and ants, experiments show that individuals approaching a landmark will often orient themselves from the same vantage point so that information about target location is assessed from the same relative position associated with the landmark. Such studies also show, however, that bees can recognize the pattern of an artificial flower even when approaching from a novel direction. When they first leave a new feeding area, they often hover around the area, presumably picking up snapshots of what the local scene looks like from different approach directions. This is useful because a feeding area can, in theory, be approached from an almost infinite variety of directions, especially when one considers both compass direction and height of approach. These observations from insects are paralleled by those from birds, rodents, monkeys, apes, and humans.

Imagine the following apocryphal situation. While a rat is traveling home from a foraging expedition, a small tornado picks him up, spins him around a few times in the air, and drops him down. How does the disoriented rat find his way home? If he is close to home, one might expect landmarks to be used. In fact, they are not. Ken Cheng, an experimental psychologist, simulated the twister experiment in 1986 by placing rats in a rectangular enclosure. One enclosure consisted of four white walls and no landmarks. A second enclosure consisted of three white walls and one black wall. Each rat watched as a corner was baited with food. For tests run in the second enclosure, the food was always placed at the junction between a white wall and a black wall, and was also associated with a distinctive smell such as licorice. The diagonally opposite corner (a junction between two white walls) was associated with a peppermint smell. To disorient the rats, an experimenter covered their eyes and slowly rotated them. After a few spins, they were released and allowed to search for the baited site. The rats' search behavior was the same in both enclosures, looking half the time in the correct corner and half the time in the diagonally opposite corner. Apparently, the rats ignored both the visual and olfactory cues, attending exclusively to the geometry of the room.

What explains the rats' pattern of search? The fact that they searched in the incorrect but geometrically equivalent corner indicates that they were reorienting by comparing the shape of the environment before disorientation with the shape afterwards. They used the geometry of the room, but ignored the nongeometrical features such as the colors of the walls and the smells associated with the two corners. Because nondisoriented rats can certainly find food on the basis of visual and olfactory cues, Cheng's experiment shows that disoriented rats use a different navigational mechanism, one that is functionally blind to landmarks. This is precisely the kind of observation that is crucial for the cognitive map hypothesis. Specifically, the rats should not make these kinds of systematic mistakes if they don't log geometrically relevant information about target locations. And geometric computation is precisely what is needed for the construction of a cognitive map.

The developmental psychologists Linda Hermer and Elizabeth Spelke ran Cheng's experiment with human adults and toddlers, using both an all-white rectangular room and one with highly noticeable landmarks. Would humans act like rats? Would adults differ from toddlers? Following disorientation in the all-white room, the adults and toddlers preferentially searched in the correct corner and the diagonally opposite, but incorrect corner. When subjects were tested in the room offering color cues, adults again preferentially searched in the geometrically appropriate corners, but searched more often in the correct than in the incorrect corner. These results show, then, that following disorientation, adults orient by reference to a map of the room's geometry and use relevant landmarks, if available, to refine their search. In contrast with adults, toddlers preferentially searched in the geometrically appropriate corners, but showed no evidence that they were aware of the highly salient color cues such as one shiny red wall bracketed by three white ones, a child-sized plastic elf in one corner, and a similarly large plastic trash can in the other. The toddlers searched equally often in both corners, as did Cheng's rats.

Once again, these results show broad taxonomic similarity in navigational systems. For both rats and human children, accurate navigation following disorientation is based on a mental tool that references stable geometric cues (e.g., the shape of an enclosure) and ignores or, more precisely, is blind to nongeometric cues such as landmarks. Although it is not yet clear why adults attend to landmarks and children don't, Hermer and Spelke have speculated that it may have something to do with the adult's capacity to use language to encode locations in space. For example, while searching for a ball in a room with one blue wall, adults might rehearse the sentence "The ball is in the corner to the left of the blue wall." If adults use language in this way, then their navigational abilities should deteriorate when the language system is used for another purpose. In a recent experiment, Spelke has shown that if adults are required to repeat a sentence out loud while attempting to find the target location, they fail. By talking out loud, adults are unable to repeat the relevant linguistic information about spatial location and therefore are unable to rehearse the planned navigational route.[16] This implies that in the absence of language, abstract spatial representations simply cannot evolve, or in the case of humans, cannot be learned. Contrary to this claim is the work showing that Clark's nutcrackers use the concept *middle* to find food. This is an abstract spatial representation, and it is tapped in the absence of language.

The navigational errors exhibited by rats and children are surprising. The landmarks are so obvious, and so obviously ignored. Such blindness to relevant information shows that the underlying system for navigation is operating on its own terms, following its own principles. Let me illustrate the level of ignorance animals achieve when they think they know what they're doing. Train rats to work their way through a maze to find a small piece of food at the end of one branch. Once they are successful at this, split the rats up into groups. For one group, a wall is inserted midway in the branch of the maze with food. As the rats turn the bend into this branch, they run smack into the wall, dropping back, stunned! For the second group, the baited branch of the maze is sawed off at its midpoint. When these rats turn the bend, they run right off the edge, their legs spinning in midair like Wile E. Coyote in an old Roadrunner cartoon. Finally, a large heap of food is placed halfway between the start of the branch and the target piece of food. These rats run right through the heap as if it was nonexistent. The rats are on a dead reckoning course to food, and the most obvious changes to the environment are ignored. Three blind mice, see how they run.

All animals are equipped with a mental tool for processing spatial information. Although animals may use different sensory modalities to navigate in space, all animals share in common the mechanism of dead reckoning. This core mechanism is then complemented by a set of specialized devices, each designed to meet the unique problems posed by the animal's environment, be they trekking through a featureless desert, a murky river, or a tropical rain forest.

WHY, WHAT, AND WHERE?

When animals travel, they are usually searching for one of three fundamental resources: food, mates, and a place to live and reproduce. Together, these three dimensions define a species' environment, its ecological space. By focusing on what animals do in this space, we can address three questions. First, *why* do animals take particular travel routes and not others? Second, *what* are they searching for and what mechanisms make such searches possible? Third, *where* are they likely to find the targets of their search and how do they remember where they are? This final section explores the three Ws of spatial knowledge, integrating insights from neurobiology and evolutionary theory.

Many species are highly promiscuous in their mating habits. Males roam over large distances to find receptive females, whereas females occupy relatively small home ranges. Sex differences in ranging behavior create differences in selective pressures. Males should have better memories than females because they have to recall the distribution of females over a large range; females simply wait for the males to come courting and then exert their preferences. But better memories require better storage devices, and for many animals, the region of the brain that is responsible for such deep storage is the hippocampus. We would therefore expect males to have a larger hippocampus than females in species where such selection pressures arise.

In the mid-1980s the anthropologist Steve Gaulin started exploring the possibility of sex differences in human spatial ability. Appreciating the difficulties associated with studying their navigational patterns as well as their brains, he turned to rodents, specifically, a polygamous meadow vole and a monogamous prairie vole. In the meadow vole, males expand their home range size during the breeding season in order to increase their mating opportunities. Monogamous prairie voles do not. When these species are taken into the laboratory, male meadow voles outcompete females in maze running, a test of spatial ability. Prairie voles, in contrast, show no sex difference in maze running. When the brains of these animals are examined, male meadow voles exhibit a relatively larger hippocampus than females, whereas prairie voles show no sex differences in hippocampal size.[17]

Gaulin's results suggest that when ecological pressures differ between males and females, selection can cause changes in the brain. Such changes are highly specific, however, limited to the particular regions of the brain that are most significantly involved in coping with the pressures imposed. How general are these findings? Did Gaulin just happen to luck out, picking two species that are extremes, outliers in a distribution? No. Researchers such as the ethologist Lucy Jacobs have demonstrated similar effects in other rodents, including Merriam's kangaroo rat. Further, studies of birds reveal almost identical patterns of change in the hippocampus as a function of ecological pressures, but the pressures are different from those documented for rodents.[18]

Clark's nutcracker is confronted with an exceptionally difficult recall problem. In the fall, individuals store twenty to thirty thousands seeds in five to six thousand locations and then wait for the spring to retrieve them, which they do with a high degree of proficiency. How can the nutcrackers pull off this incredible feat? The first insight into this problem emerged from two comparative studies looking at the relationship between the size of the hippocampus and whether the species stores its food for future retrieval. Overwhelmingly, hippocampal size was significantly larger in storing than in nonstoring species even when closely related species were compared.[19]

Studies of food storing provide an elegant illustration of how comparative analyses inform our understanding of brain-behavior relationships. They also raise a fundamental point about the evolution of psychological adaptations and for that matter, any adaptation. Specifically, we can look for psychological adaptations in two ways. First, we can study closely related species and look for differences in their psychologies. For example, among the members of the *Paridae*, a songbird family, the great tit represents an oddball species in that it is one of the few that fails to store food. Apparently as a result, the hippocampus of the great tit is relatively smaller than in all the other food-storing species in its family. Second, we can compare distantly related species that share the same psychological adaptation. For example, although chickadees, nuthatches, and jays are only distantly related, all three species store their food, and all three show the same pattern of increased hippocampal size and improved spatial abilities over closely related species that are not storers.

Given the overall pattern revealed by comparative studies, what can be said about the specifics of storing, spatial ability, and the hippocampus? It turns out that the answer converges with the work we have discussed earlier. The hippocampus plays a functional role in the process of food storage and retrieval. Black-capped chickadees store food and show the characteristic hippocampal enlargement. If you remove the hippocampus from adult birds, they store, search, and eat as much as intact birds. However, they rarely retrieve food from the sites they filled. They fail to remember where they stored the goods. Furthermore, whereas chickadees without a hippocampus can find food based on the color and shape of landmarks, they are incapable of finding food on the basis of spatial position alone. This emphasizes the point that the hippocampus is involved in spatial memory.

All the work described thus far shows how animals with mature brains process spatial information. But how does the young bird acquire spatial knowledge? Is the hippocampus relatively large at birth, prior to the experience of storing and retrieving? Recall that the work on indigo buntings revealed a critical window for experiencing the stellar configurations. If the young bird's view of a patch of sky was occluded during the fledgling stage, it failed to orient in the correct direction during its initial migration. Is there a parallel story for storing and retrieving? Yes, but the window of opportunity may be wider. In 1994 the biologists Nicki Clayton and John Krebs raised a group of marsh tits in an environment where they were fed but never allowed to store and retrieve seeds; marsh tits naturally store food in the wild. Other individuals obtained such experience, but at different levels and at different ages. When compared on spatial tasks, individuals with experience paid more attention to location cues over color/shape cues, whereas individuals without such experience paid equal attention to both cues; the amount of experience with storing and retrieving had no effect on spatial ability. Moreover, individuals without experience of storing and retrieving showed a decrease in hippocampal volume and an increase in the number of dead neurons; as long as there was some experience during the fledgling period, the hippocampus developed normally.

Clayton and Krebs conclude that experience is fundamental to hippocampal growth, but the window of opportunity appears broad when compared with other developmental studies that explore the importance of sensitive periods in development (see chapter 6). The influence of experience on the brain is specific, showing its effect in the hippocampus rather than the entire brain. Moreover, variation in the size of the hippocampus between species appears to reflect an anatomical adaptation for food storing.

The Clayton-Krebs conclusions fall squarely within the general theme that we have been developing in the previous chapters, specifically, the difference between domain-specific and domain-general learning devices. But how robust are they? If marsh tits are given experience in a spatial task that involves memory, but not the experience of storing and retrieving, then hippocampal growth is normal. Thus, the functional enlargement of the hippocampus may not depend on specific kinds of experience, something that a strict domain-specific position requires. Furthermore, although the size of the hippocampus is clearly variable, showing changes in response to foraging demands, it also appears to be responsive to other sorts of pressures. Gaulin and Jacobs's work on rodents indicates that sexual pressures can cause changes in hippocampal volume. More recently, David Sherry and his colleagues have demonstrated that sex differences in parental care can also exert pressure on hippocampal function. In some birds, individuals dump eggs into the nests of other species and thus relieve themselves of the burden of parental care. These brood parasites have different strategies. In some species, the male and female travel together to find target hosts, whereas in other species, the female carries the sole responsibility of finding a host, and often this involves the recall of dozens of nest sites. In species where the female is responsible for finding a host, her hippocampus is significantly larger than the male's; in species where the female and male search together, there are no differences in hippocampal volume.[20] These observations suggest that the spatial learning mechanism may be more general than originally conceived; we may have to revise our views about domain-specificity, at least in the domain of spatial navigation.

All species carry specialized mental tools for processing information about objects, number, and space. For some species, the relevant objects are fruits, the relevant numbers are three or less, and the relevant space is the distance between fruiting trees. For other species, the relevant objects are predators and other group members, the relevant numbers are one, two, three, or more, and the relevant space is a tropical rain forest filled with competitors. Such variation leads to differences between species in the kinds of mental devices added to each domain of knowledge, devices such as a cognitive map, and a count system that is based on a limitless set of symbols. How then did some organisms evolve into nature's psychologists, individuals with a sense of self, a capacity to imitate, and the ability to deceive?

PART II

Nature's Psychologists

All animals have a mechanism to recognize others, but only a small number of animals have the capacity for self-recognition.

5

Know Thyself

In mythology, Narcissus is a handsome yet conceited young man who falls deeply in love with the nymph Echo. Rather than fall for his romantic overtures, Echo simply repeats Narcissus's words back to him. Narcissus considers this disrespectful. Because of his egotism and social ineptness, the gods curse him. The curse comes to fruition. Upon seeing his reflection in a forest pool, Narcissus falls madly in love with the image, ignoring all else in the world. Ultimately he wastes away.

Like the pool of water, mirrors also reflect our image. Perhaps because of their seemingly magical properties, they have played a key role in many stories. Tigger, Winnie the Pooh's friend, is surprised to find that there are other Tiggers on earth when, for the first time, he stares into a mirror and fails to recognize himself. In his book *The Words*, the philosopher Jean-Paul Sartre criticizes culture, "a product of man: he projects himself into it, he recognizes himself in it; that critical mirror alone offers him his image." But the mirror in "Snow White and the Seven Dwarfs" is the most magical of them all, for it has the power to see into the future and give an honest assessment of reality. Let me capture the crucial bit, crucial because of the assumptions being made about the reader's beliefs.

The queen of the castle is wicked and vain. Every day she stands in front of her mirror waiting for the usual answer to her rhetorical questions about beauty. But now think about the cognitive steps required for the reader to appreciate those famous words: "Mirror, mirror on the wall, who is the fairest of them all?" The first step is to see that the queen wants the mirror to render a verdict on her beauty. For this to work, the queen must appreciate her own beauty relative to that of others. She thereby must have some understanding of self—a subjective view of herself as an individual in the world, distinct from other individuals. Second, the queen thinks that the mirror has a reasonable understanding of beauty and uses such intuition to evaluate the beauties in the land. Third, when the mirror states that the queen is the fairest of them all, she is satisfied, because this of course matches her own beliefs. One day the mirror proclaims Snow White the fairest. Not only is the queen furious, driven by jealousy, she is enraged to discover that she had harbored a false belief. The magic in this story, therefore, can be appreciated only if these steps can be followed. And at the root of each step is an understanding of self-awareness, an appreciation of who we are and how we come to know our inner selves, what we believe, desire, and want.[1]

How do humans acquire an understanding of self and why did this capacity evolve? Are infants born with some rudimentary cognizance of who they are? Do any animals share our capacity for self-knowledge, and if so, what form does their representation take?

KINDS OF RECOGNITION

Consider the circumstances in which it might pay for some members of your species to be treated differently than others. Evolutionary biologists have a clear answer to this problem: whenever discrimination leads to fitness payoffs—more surviving offspring and therefore more genes passed on to the next generation. In sexually reproducing species, one criterion for discrimination would surely be the sex of the individual. Attempting to mate with the same sex is a dead end. Mating with a sexually immature individual is equally unproductive. Thus, a distinction between juvenile and adult is also needed. Kin must also be discriminated from nonkin. Mating with close relatives is generally deleterious, yielding offspring with lower chances of survival than those produced by mating with distant kin or genetically unrelated individuals. On the other hand, helping close kin survive and reproduce—an act of biological altruism where there is a cost to the actor and a benefit to the recipient—is advantageous. Because kin share genes in common, helping relatives indirectly helps the replication of the altruist's genes.

Selection therefore favors individuals that can discriminate among others on the basis of their degree of genetic kinship. Sure enough, all species are capable of some level of kin discrimination, although the mechanisms remain a mystery for many animals.

All social, sexually reproducing organisms seem to be equipped with neural machinery for discriminating males from females, juveniles from adults, and relatives from nonrelatives. What kinds of mental tools have evolved to facilitate discrimination among individuals that differ with respect to genetic relatedness? All recognition systems must find a balance between specificity, rigidity, and accuracy on the one hand, and generality, plasticity, and inaccuracy on the other.

Nature provides us with several cases where a simple familiarity rule guides the preferential treatment of kin: those who you encounter often are your kin. A classic example of this kind of rule in action is imprinting, the pattern of attachment observed in many birds. As Konrad Lorenz, the Nobel laureate in ethology, observed in the 1940s and 1950s, young chicks tend to follow the first object that they see moving. This is as good a rule as any, given that in nature, the first object tends to be Mom. Devious experimenters, however, have taken advantage of this rule to demonstrate that the young chick will follow (i.e., imprint on) a scientist, a rotating red cube, or even detached chicken parts that have been plastered onto a box, Jackson Pollock style. Such following behavior is not a sign of stupidity, but rather the signature of a domain-specific mechanism that generates intelligent errors, akin to those we encountered in the last chapter on spatial navigation.[2]

In contrast to a familiarity rule, some recognition mechanisms are like templates, blueprints for discrimination. They are highly specific, relatively rigid, and accurate. The tunicate, a marine invertebrate, settles next to and sometimes fuses with individuals who share the same immune system genes. Sweat bees have colony-specific odors, determined by genes coding for the structure of pheromones, a kind of signature perfume. By identifying how someone smells, they can reject intruders. In ground squirrels, individuals that have been reared apart nonetheless prefer to associate with unfamiliar siblings over unfamiliar nonsiblings.

The evolutionary biologist Richard Dawkins coined the phrase "green beard effect" in the 1970s to describe how some of these systems might work in the service of identifying like-minded altruists—individuals willing to incur a direct fitness cost so that another individual may benefit. In general terms, if I have genes that predispose me to behave altruistically, I want to find someone who is similarly predisposed. I can't see their genes, but I can see their bodies, smell their body odors, hear their voices, and watch their actions. Such expressions should provide me with a reliable clue. If I have a green beard and they have a green beard, then chances are we share genes in common. Consequently, selection will favor altruistic actions toward green beards over red beards or clean-shaven individuals.

A rigid template, or lock and key mechanism for recognition, is prone to errors of a different kind from those exhibited by more malleable mechanisms. Specifically, relatively minor changes in the orientation or configuration of the target object could result in false recognition—thinking it's a mate when in fact it's a predator. Imagine that the template is set up to recognize your brother John by specifying what John's face looks like in an upright position. The system cranks through all the relevant features: Big nose, check! Blue eyes, check! Broad chin, check! Bushy eyebrows, check! Yes, it's John. You turn your back on John for a few minutes and when you look around again, he's standing on his head and you can't recognize him. Nothing matches exactly, except the color of his eyes. The template is too rigid. Either you have to tilt your head or you have to wait for the person standing on his head to go upright.

Dawkins's green beard effect provides an explanation for why individuals that look, smell, or sound alike help each other. The green beard effect can also be read in a way that implies some level of self-recognition. In particular, to know that your green beard is like mine, I have to know what *I* look like. The slippery concept in this last sentence is "to know." Perhaps the simplest sense of knowing, and certainly the most uninteresting from the perspective of the mind, comes from our understanding of immune systems. A molecule entering a system is considered compatible if and only if it matches a particular molecular configuration or signature. Forging the signature is, of course, possible and amply demonstrated by the deadly sophistication of viruses and parasites.[3] The immune system's "I" is, therefore, a molecular signature, and "knowing" translates to "recognizing." There is no sense in which the immune system is *aware* of itself, aware of what its signature looks like. There is no reason for it to know in this sense.

There are other recognition systems that may be like the immune system in that they're equipped to recognize signals with a distinctive signature, but are equipped with more sophisticated machinery. Consider birdsong (see chapters 6 and 8), and in particular the process of song acquisition in a species such as the white-crowned sparrow that learns only one song during its lifetime. During development, the young bird progresses through a series of stages before it produces crystallized song in adulthood. In stage one, the song material is processed and stored in memory. The material stored depends on several factors, an important one being whether or not there is a match between the acoustic morphology of the input and the morphology of the template stored in the brain. In a species like the white-crowned sparrow, if the individual hears material other than white-crowned sparrow song, and hears it outside a specific window of opportunity for learning, the bird produces a song that shares little in common with its species' signature song. If all goes well, however, the bird stores the material in memory and then, at a later date, produces first a sloppy rendition and then a virtuoso performance.

Does a white-crowned sparrow with a crystallized song distinguish this song from one produced by another bird? How might we find out? The answer lies in some fancy neurobiology and some simple playback experiments. In 1983 the neuroscientist Dan Margoliash anesthetized white-crowned sparrows that had acquired their species-typical song. While they were asleep, he played back different acoustic signals and recorded neural responses from a part of the brain involved in song processing. The signal eliciting the strongest neural response was the bird's own song. If the subject heard a white-crown song from a different population dialect, the neurons fired weakly. If the subject heard a white-crown song from the same population dialect, the neurons fired more. If the subject heard its own song played backwards, the neurons fired, but not as much as when its own song was played forwards. These neurons, therefore, appear to be tuned to one station—the bird's own song. If the brain codes the song in this fashion, what happens at a higher perceptual level, the level at which the bird must listen, process, evaluate, and respond to a song?

When birds hear an unfamiliar intruder in their territory or near their boundaries, they typically approach and respond aggressively. When they hear their neighbor singing, they might counter-sing, but typically do not respond aggressively. This suggests that they make a distinction between familiar neighbors and unfamiliar intruders. Does the bird classify its own song as an intruder, a neighbor, or something else? According to playback experiments of recorded songs, response to strangers is most intense and to familiar neighbors least intense. Self-song elicits an intermediate response, suggesting that it is perceptually different from a neighbor and a stranger. But can we say more? Unfortunately not, and here the sensitivity of our measuring techniques limits our understanding. Birds may respond with intermediate levels of aggression to self-song because they have heard it more often than a stranger's song, but either less than or as often as the familiar neighbor's song. However, since it is clearly not a familiar neighbor, it elicits a weaker response than the stranger's song. As we all know, hearing our own voice played back to us is a bizarre experience—it doesn't really sound like us.[4]

ON REFLECTION

The kind of self-recognition we have discussed thus far is simple. It is grounded in measurable features of the external world, and restricted to a single modality—recognizing molecules, odors, and sounds. But our own understanding of self—of who we are and what we believe—extends beyond the physical world. It is also multi-modal. We don't just have a sense of what we look like, what we sound like, and what we feel like. Our sense of self integrates all of these. We can reflect on what we believe and how our beliefs are sometimes similar and sometimes different from the beliefs of others.

Normal adults living in the twentieth century are frequently reminded of their appearance. We see ourselves in the mirror, in television monitors that capture our actions as we walk into a department store, and in photographs. Such physical representations provide an image of what we look like, an image that with time becomes indelibly marked in our minds. Although we all tend to think that our mirror reflections and photographic portraits distort what we really look like, the bottom line is that we have a good sense of our external appearance; our capacity to recognize self by means of our other senses is far less accurate than it is in the visual domain. We all know about the things that can change our physical appearance: makeup, distorting mirrors, tinted contact lenses, masks, hats, hair dye, and plastic surgery. When I say "know," I am referring to a capacity to recognize change, reflect on it, and understand how our appearances change and what such changes do with respect to who we are and what we think. When I say "we," however, I am referring to a special class of organisms: normal humans, the world over, who are at least two years old.

When a one-year-old looks into a mirror after a small odorless mark has been surreptitiously placed on her forehead, she fails to notice a change. She does not, for example, selectively touch the mark. There is little evidence that children at this age recognize themselves in the mirror. Somewhere around the second birthday, the child will touch the mark on her face. She recognizes her mirror image, a source of joy and entertainment.

Joining the one-year-olds are two other groups of humans who also have problems with familiar faces including, sometimes, their own—patients with neurological problems known as *prosopagnosia* and *Capgras syndrome*. Among prosopagnosics who are unable to recognize familiar faces, neurological analyses reveal damage to areas of the brain involved in complex visual processing (inferior temporal lobes). Although such patients fail to recognize the faces of famous as well as familiar people, they can recognize them by voice, and generally have no problem recognizing other objects. But the most fascinating, and psychologically devastating, problem for a subset of prosopagnosics is the inability to recognize their own face— whether in a mirror reflection or in a Polaroid picture. For these individuals, the mirror image simply reflects another individual.[5]

Research on both normal subjects and patient populations indicates that individuals are often unaware of the knowledge stored in their brain. In the 1980s the neuroscientist Antonio Damasio and his colleagues set out to determine whether prosopagnosic patients might have some *covert* recognition of familiar faces, a kind of recognition that they may be unaware of. To capture the intuition, imagine that you are the president of the United States and you meet a group of supporters in a small rural community. The people line up on Main Street, waiting to shake your hand. You walk along, shaking each Tom, Dick, Harry, Mary, and Jane's hand. They give you a warm greeting. Harry has a joy buzzer in his hand, a gag that gives you a small shock. You feel the mild shock, look puzzled, but then laugh and move on to the next supporter. Now, a few days later, you are in the same town and approach a group of people. Although you don't recognize anyone's face, when you see Harry your heart starts racing and you begin sweating. You detect something familiar, but you can't quite put your finger on it. You don't remember the joy buzzer, but seeing Harry gives you a feeling of discomfort. Damasio and his colleagues ran this experiment with prosopagnosic patients. What they found was that although the patients couldn't tell you who was familiar and who wasn't, they experienced an increase in heart rate and the sweatiness of their skin when they saw the face of the person who shocked them.

To elicit some feeling of familiarity for a prosopagnosic, you have to change the emotional significance of the face. But boosting the emotional signal isn't sufficient to bring about recognition of a face. The prosopagnosic patient will never be able to tell you, "That's Harry." They only have a hunch that can't be articulated as to who Harry is. Although this experiment has not, to my knowledge, been run while a prosopagnosic patient sees a picture of him or herself, the same kind of effect might emerge: faster heart rate and sweaty palms, but sadly, not a "That's me!" response.

Capgras syndrome differs from prosopagnosia in that patients recognize familiar individuals, but think that such individuals have been replaced by imposters. When shown a picture of their mother, they will correctly identify her, but will also say that she has experienced something akin to an alien abduction. Unlike prosopagnosics, Capgras patients show no difference in the sweatiness of their skin or heart rate when presented with familiar and unfamiliar faces. These observations suggest that the areas of the brain that process faces (inferior temporal lobe) have been disconnected from the areas of the brain that provide faces and other objects with emotional significance (amygdala). For normal people, faces are associated with feelings—Mother Teresa with warmth and compassion, Hitler with hatred. In one fascinating case, a Capgras patient with the initials D.S. thought of himself as a double, and was able to articulate why others failed to recognize him, stating that it was because "I'm not the real D.S." In speaking with his mother, he said, "Mother, if the real D.S. ever returns do you promise that you will still treat me as a friend and love me?" Like prosopagnosics, then, Capgras patients also experience a disruption of self, but one that is more centrally connected to the subjective properties of self, what it is like to have private thoughts and emotions.[6]

What about self-recognition in animals? Are animals like normal human adults, human infants, some prosopagnosics, or Capgras patients? If some animals are like normal human adults and others are like human infants, then what accounts for such variation? Not surprisingly, perhaps, Darwin provided some answers to these questions. In considering some of his own observations, he remarked, "can we feel sure that an old dog with an excellent memory and some power of imagination, as shewn by his dreams, never reflects on his past pleasures in the chase? And this would be a form of self-consciousness."[7] Crucially, not only was Darwin's thinking advanced, he also tested his ideas. In the 1870s he conducted the first mirror test for self-recognition.

> Many years ago, in the Zoological Gardens, I placed a looking-glass on the floor before two young orangs, who, as far as it was known, had never before seen one. At first they gazed at their own images with the most steady surprise, and often changed their point of view. They then approached close and protruded their lips towards the image, as if to kiss it, in exactly the same manner as they had previously done towards each other, when first placed, a few days before, in the same room. They next made all sorts of grimaces, and put themselves in various attitudes before the mirror; they pressed and rubbed the surface; they placed their hands at different distances behind it; looked behind it; and finally seemed almost frightened, stared a little, became cross, and refused to look any longer.[8]

What do we learn from Darwin's observations? What were the orangutans thinking as they looked into the mirror? Let's accept Darwin's assumption that these particular orangutans had never seen themselves in a mirror. From this premise, we have a set of observations that unfold over time. Initially, the orangutans appear slightly alarmed and curious about the reflection. Then, like all creatures confronting novelty, they approach with caution. Their response is friendly, a few kisses to the mirrored glass. Next, they produce grimaces, a kind of smile, while touching the mirror and looking behind it. In terms of the orangutans' thoughts and feelings, these responses are somewhat ambiguous. Grimaces can be friendly or fearful. In both cases, however, it would appear that they are communicating with one another. An alternative interpretation, however, is that the orangutans are taking mental notes on what their own expressions look like. The former suggests that they see the image as someone else; the latter suggests that they see it as themselves. The same pairing of explanations holds for their searching response. It may be that they are looking for someone else behind the mirror, or they may be trying to figure out how the mirror works. We simply don't have enough information to decide one way or the other. When they stared and became "cross," their emotional state changed from relatively friendly to aggressive. But again, aggressive expressions need not imply that they perceive the reflection as a threatening stranger. Similarly, their loss of interest is equally uninformative.

Though Darwin was well ahead of his contemporaries, this particular experiment leaves us dangling in uncertainty. We simply can't say, one way or the other, whether orangutans are capable of recognizing their image in the mirror.

In 1970 the comparative psychologist Gordon Gallup reopened the question of self-recognition in animals and developed an elegant yet simple test to investigate it. He started by presenting a fixed, full standing mirror to an adult chimpanzee and then watched its response—Darwin's test. At first, the chimpanzee stared at, touched, displayed into, and explored behind the mirror. Over time, these social behaviors waned as the chimpanzee began to use the mirror to look at previously unseen parts of the body (e.g., inside the mouth). Such use of the mirror suggests that the chimpanzee perceives the reflection as an image of self, whereas displays and exploratory behavior can be interpreted either in the same way, or as evidence that they perceive another chimpanzee. To provide a more quantitative analysis of their behavior and to test between the alternative interpretations, Gallup anesthetized the same chimpanzees. While the chimps were unconscious, he placed one red mark on their right eyebrow and one red mark above their left ear; there were no tactile or olfactory cues associated with the red marks. When the anesthesia wore off, observations were recorded both with mirror present and mirror absent.

With the mirror in place, chimpanzees spontaneously and immediately touched the red marks on their eyebrow and ear; before the mirror was uncovered, the number of touches to the marked areas were no different from touches to the unmarked areas, and individuals with no prior exposure to mirrors never touched the marks. After touching these areas, some chimpanzees looked at their fingers; one individual both looked at and sniffed its fingers.

What kinds of emotion did the chimpanzees experience upon first recognizing their mirror reflection? Did they have new thoughts about their physique? Although Gallup did not assess whether such knowledge—whatever it might be—had an effect on the chimpanzees subsequent behavior, it would be fascinating to understand how information about one's appearance ties into one's thoughts and actions. Is there an increase in attention to appearances? Is a chimpanzee more self-conscious of its actions and how they appear to others? How does the visual representation of self impact on the more general, multisensory representation of self?

Gallup's results were important, as was his method. It provided the kind of experimental technique that could be used across species with little to no modification, thus providing a comparative measure of cognitive capacity. But what capacity was being measured? For Gallup, the mirror test provided evidence not only that chimpanzees recognized themselves in the mirror, but also that they were self-aware.

In our discussion of "Snow White and the Seven Dwarfs," I suggested that to understand the evil queen's response to the mirror, one must make several cognitive steps. The queen not only recognizes herself in the mirror, but also knows what she believes about her own beauty. If she was not aware of her own beliefs, then when the mirror states that she has lost the beauty contest to Snow White, her ensuing rage would make little sense. In short, a self-aware individual is cognizant of his or her beliefs, how they change, and how they differ from the beliefs of others. Gallup believes that the mirror test indicates that chimpanzees are self-aware. I don't think the mirror test provides any leverage with respect to self-awareness. If it did, we would have to conclude that prosopagnosic patients who fail to recognize their mirror image have no understanding of their beliefs. However, it is precisely because prosopagnosic patients have self-awareness that their inability to recognize a mirror reflection is so painful. Before we enter this controversy, let's review how other animals respond to their mirror reflection.

On the heels of Gallup's original publication, scientists conducted mirror experiments with both closely related species—other apes—and more distantly related ones—pigeon, parrot, elephant, dolphin, and several monkey species. In a number of these experiments, all the steps in Gallup's original procedure were followed, so the results were more directly comparable. Two great apes, the orangutan and the bonobo, reacted like chimpanzees to the mirror, selectively touching the marked areas, but not the symmetrically matched but unmarked areas. Of more than twenty-five gorillas tested, only one has responded like these other great apes, and this individual is no ordinary, chest-banging, celery-crunching gorilla. Rather, the gorilla is Koko, the star of Penny Patterson's research project. Over the course of several days, an experimenter patted Koko on the head with a damp cloth while another experimenter carried a large mirror in front of her. Koko showed no interest in the mirror. One day, the experimenter patted Koko on the head with a nontoxic, odorless color mark. When she saw herself in the mirror, she did a double take and wiped off the mark while looking in the mirror! This intriguing observation is complemented by another tantalizing one: while Koko was looking into the mirror, a trainer asked what she saw. She signed, "Me, Koko."[9] Why Koko appears to show mirror self-recognition, whereas all other gorillas don't, remains a puzzle. Perhaps there is something special about being reared by humans, like a human, and in a human environment?

Until 1994 all monkeys (e.g., rhesus, capuchins, marmosets) tested on Gallup's procedure failed to show mirror self-recognition. They either showed no self-directed behavior or responded aggressively to their mirror reflection. Furthermore, although these animals failed Gallup's test, they were able to use a mirror to find hidden objects. For example, rhesus monkeys discriminate objects such as snakes and food when presented as mirror reflections, moving away from the snake reflection and approaching the food. Marmoset monkeys and fish can do this as well.[10] This observation is important because it shows that these species grasp at least one aspect of the mirror's reflective properties, though they fail to notice that this particular property of mirrors holds for all objects, including the self.

Most discussions of this work conclude that the great apes and human children over the age of two years have some understanding of self, but no other species do. Thus, an evolutionary gap separates humans and great apes from all the other creatures. If correct, this is a major finding, for it sets the stage for an evolutionary scenario about the origins of self-knowledge that has relatively recent roots in the tree of life. How confident are we, then, in the current species divide? Are there are other tests or modified versions of Gallup's original design that tell another story?

If you were trained in the Skinnerian tradition, you would not accept the claim that animals have beliefs, desires, and a sense of self. You would argue instead that what looks like behavior guided by beliefs is actually behavior shaped by a history of reinforcement. When a chimpanzee touches its eyebrow while looking in the mirror, this has nothing to do with a sense of self. Rather, the chimpanzee touches itself because it has been shaped by consecutive episodes of reinforcement.

In 1981 the psychologist Robert Epstein, one of Skinner's last students, set out to determine whether pigeons are capable of using a mirror to touch selectively a mark placed on a previously unseen body part. Epstein hoped to lay to rest the idea that using a mirror to touch oneself has anything to do with either self-recognition or self-awareness.

Using four, mirror-naïve pigeons, Epstein trained these individuals first to peck blue dots on a wall. Next he trained them to look in a mirror to see a blue dot flashed on the wall behind. If they turned and pecked the blue dot, they received food. Finally, Epstein placed a white collar around their neck and a blue dot on their chest; because of the collar, the pigeons were not able to see the blue dot on their own. When collar and dot were in place, Epstein uncovered a mirror. The pigeons turned, looked in the mirror, and pecked the blue dot.

What can we conclude from the pigeon exercise? Epstein argued that pigeons can use a mirror to locate an object, including one placed on their own body. Further, such behavior does not justify the conclusion that pigeons have "self-awareness" or a "self-concept." Consequently, terms suggesting a comparable mental state should not be attributed to the chimpanzee, for similar responses may have been trained as well.[11]

The key problem with Epstein's conclusion is that the pigeons were trained, through a complicated regime of reinforcement, to use mirrors and to peck at blue dots. Recall that in Gallup's original report (and all others that have followed), he first observed chimpanzees' *spontaneous* responses to mirrors. Working from these observations, which already suggested that they recognized their own reflection, he conducted the mark test. In no case was the chimpanzee rewarded for its actions. Epstein's experiment therefore shows that pigeons can use a mirror to identify spots on their bodies, but only through a regime of reinforcement. Chimpanzees engage in self-directed behavior in the absence of training. In both cases, we can conclude that individuals understand some aspects of a mirror's reflective properties. But Epstein's results do not exclude the possibility that what the chimpanzee recognizes is an image of itself in the mirror.

In those species that appear to recognize their mirror image, do all individuals in the species respond to the mirror in the same way? If there is variation within the species, what are the underlying causes? In a test of eleven chimpanzees, where the amount of exposure to mirrors was varied prior to the dye mark test, only one chimpanzee showed evidence of mirror self-recognition. In an experiment involving a larger sample of chimpanzees, individuals under the age of eight years failed to show mirror self-recognition. In other experiments, however, individuals from two to six years recognized their mirror image, whereas some older individuals did not.

Mirror experiments with chimpanzees reveal considerable variation between individuals. At present, we don't understand whether this variation is due to differences in experimental procedures or true individual differences. In cases where young animals fail to recognize their mirror image, one is tempted to conclude that this capacity has not yet developed. If this interpretation is correct, then young chimpanzees are like human children under the age of two years. Unfortunately, however, there are no precise measurements for comparing human and chimpanzee cognitive development or brain maturation. If we take a crude measure of brain growth, a twelve-month-old human is about equivalent to a chimpanzee at birth. Brain growth tends to slow down dramatically at around three years of age for chimpanzees but not until six years for humans. Without an understanding of how different areas of the brain change during development, and how the neural connections between areas change over time, we cannot determine whether a young chimpanzee is more or less advanced than a child of a given age. Nevertheless, even if we can explain the young chimpanzee's failure as a case of developmental immaturity, the failure of some older chimpanzees to pass the test is more problematic. Since these individuals were by no means senile, or even reproductively over the hill, their failure is difficult to explain.[12]

The studies conducted to date indicate that when Gallup's original dye mark test is administered, adult great apes and human children over the age of about two years *generally* pass. No other species even comes close to showing such spontaneous behavior toward their mirror reflection. A conservative interpretation of these results posits a gap in the evolutionary tree, with one branch or class of animals understanding that the mirror reflects their own image, and a second branch failing to understand this relationship. Leaving aside for the moment the validity of this interpretation of the mirror response, we can ask another question: is Gallup's test equally *fair* for all species? Is it truly a test that can be used across species, with no modification? If subjects fail the test, how might we explain such failures?

Have you ever noticed how difficult it is to discern the actual movement of your arms and hands when you try to touch a spot on the back of your head while looking in the mirror? Or how difficult it is to determine which way you are actually moving when you look at a video monitor capturing your every action in a department store? It is hard, but one eventually solves the problem. In 1974 the psychologist Emil Menzel and his colleagues designed an elegant experiment for chimpanzees and rhesus with these difficulties in mind.

Menzel and his colleagues tested two chimpanzees, Sherman and Austin, who had participated in several language-relevant experiments (see chapter 8); they also tested a few experimentally naïve rhesus monkeys. A subject was placed in a room with a small hole in one wall. With the hole temporarily covered up, an experimenter placed a small piece of fruit on one of several possible grid squares on the back side of the wall. By sticking its arm through the hole while looking at the mirror reflection, the subject could navigate to the appropriate place and retrieve the fruit. The chimpanzees passed on the first go, looking at the reflection to find the fruit. Rhesus monkeys failed again and again. In fact, when rhesus saw an image of their hand approaching the fruit, they produced threat vocalizations and facial expressions as if it was the hand of a competitor.

Following the chimpanzees' success, Menzel and colleagues removed the mirror and replaced it with a video camera aimed at the grid on the wall. Then they connected a TV monitor to the camera and stationed it in front of the chimpanzee. The task: look at the monitor's image of your arm and the wall to find the fruit. The chimpanzees did this immediately. To eliminate the possibility that the chimpanzees were using touch to find the fruit, a black ink spot was placed in one of the grid squares; they were given a food reward by the experimenter after touching or moving close to the spot. The chimpanzees solved the problem, touching the ink spot and then withdrawing their hands to look at and sniff the black smudge on their fingers. In fact, they found the spot when the monitor revealed either a true representation of their actions, a mirror image, a 180-degree inverted image, or an image that was reversed and flipped upside down. My guess is that not many humans could match the chimpanzees' competence in this task.

Do these results show that chimpanzees recognize their own arm in the monitor? Were they thinking something like, "That's my arm on TV. And there's the spot. If I watch my arm in the monitor, I can find the spot"? Not quite. Clearly, the chimpanzees readily located the ink spot by watching the monitor's image. But they may have also used the image on the monitor as a tool that provides information about where the spot is. By moving its arm, the chimpanzee causes a change in the image. Eventually, the movement of its arm enables it to locate the black spot and thus reach for it. In this description, the chimpanzee does not have to recognize that the arm moving is "my arm." Menzel and colleagues cracked this problem by running three crucial controls. Instead of one monitor, they presented two. On a given trial, one monitor always showed the chimpanzee's arm in real time, and the other showed commercial television, a videotape of an earlier session, or an image of the grid system with the black spot, but no arm. Although we don't know what kind of preferences chimpanzees have for television, whether they would prefer an episode of *Wild Kingdom*, *Star Trek*, or *The Simpsons*, they did watch intently. They did not, however, reach through the hole and search while watching commercial television, and did so on only a few trials with the pretaped experimental session and the image of the spot on the wall. Apparently, after the chimpanzees realized that their own actions were unrelated to the image on the monitor, they stopped reaching through the hole. And just to be absolutely sure about the interpretation—that the chimpanzees see the image in the monitor as "My arm"—experiments either turned the monitor off after the subject reached through the hole or had the monitor present a static image of the grid. In both cases, the chimpanzees stopped reaching. These experiments provide strong support for the idea that chimpanzees recognize two-dimensional images of themselves.[13] Like the mirror experiments, however, this work does not help us understand chimpanzee self-awareness. We don't know what the chimpanzee thinks about himself and his arm, if he is thinking anything that profound at all.

Why are the responses of the apes and young children to these tasks so different from those of monkeys and other nonprimate species? To understand mirrors, you must be willing to devote some time to looking at your reflection, testing the reflective waters, so to speak. Suppose that the first time you look into a mirror you think it is someone else, a reasonable assumption because you have no concept of what you look like. If the reflection is perceived as someone else, then this someone else is staring back at you. Staring can be rude, but it can also be a sign of friendly interest, at least for some species. Humans make eye contact all the time. Spend a few minutes watching parents and their children, teachers and students, and lovers. Apes also stare at each other in a friendly way. But monkeys generally avoid it. When monkeys stare at one another, they are typically engaged in some kind of aggression. Put simply, whereas staring *can* be used as a threat among the apes, it is *typically* used as a threat in monkey species.

If staring is a threat, and the mirror reflection is perceived as a competitor staring back, then looking into a mirror is most likely aversive, negative, and upsetting, and monkeys will avoid doing it. Thus, monkeys should have great difficulty picking up on the most significant information from the mirror: the individual in the mirror matches, one for one, your every action. Unlike the more generic body sense that individuals might have when they navigate or when they groom themselves, mirror recognition has an important timing component; every motion happens simultaneously with the mirror's reflection. Now even if monkeys fail to understand that another individual can't match your every move in time and space, they should at least experience this as a highly novel interaction. The novelty of the interaction might intrigue the apes and young children, but apparently not the monkeys. They either become bored with the image or continue treating it as an aggressive competitor. What we need, therefore, is a task that forces them to pay closer attention—a task that overrides the initially aversive experience of staring at someone who is staring back.

Cotton-top tamarins are small arboreal monkeys with a spectacular tuft of white hair on top of their head. Although they had never been tested on the traditional Gallup mark test, closely related species had, and failed. During a lab meeting with my students, we started a discussion about the problem of self-recognition and the mirror test. The upshot of this discussion was simple: we were not satisfied with the current state of affairs since our species was deemed self-less by the mirror test. We wondered whether a more dramatic change than a mark above the eyebrow or ear might induce a stronger effect, an effect that would anchor their attention toward the mirror. Their punk white hair was such a temptation. What to do?

We decided to run the same sort of mark test, but instead of a mark on the brow or ear, we would dye their hair, the entire white tuft, using Manic Panic hair dye—flamingo pink, lagoon blue, apple green. Would mirror-guided behavior change if the relative salience of the marked area changed?

Like many other species of tamarin, the cotton-top has a dark black face. Thus, marks on the face are not that noticeable, although they are clearly discernible to us. In striking contrast to humans and the great apes, tamarins have virtually no facial expressions. Unlike the other tamarin species, however, only the cotton-top has a large white tuft of hair on its head. Therefore, we proposed that by changing the color of their hair, we would not only impose a relatively larger change than in most earlier experiments, but would also change a species-specific trait—a kind of signature hairdo.

One group of tamarins received exposure to the mirrors before the experimental manipulations started. Like other monkeys, the tamarins looked into the mirror, touched it, looked behind, and displayed. As with many of the great apes, this pre-experimental period was ambiguous with respect to the tamarins' understanding of the mirror. It was also substantially less impressive than what others have reported for chimpanzees. This isn't surprising, given that chimpanzees are more interested in the face and with facial expression than are tamarins.

When tamarins with prior mirror exposure woke up from the anesthesia, and from a surreptitious change in hair color, most of them touched their hair at least once while looking into the mirror. Tamarins with dyed hair who had no prior mirror exposure did not touch their tuft. Moreover, no tamarin touched marks on the eyebrow and ear, nor did they touch their hair when a white version of Manic Panic was applied. This last control shows that although Manic Panic—made of 5 percent vegetable dye and 95 percent conditioner—changes the texture of an animal's hair, tamarins touch their hair only when the color has been altered.

The experiment also generated a second relevant observation. Before their hair was dyed, bouts of staring into the mirror were short. Once the tamarins' hair was dyed, however, they sat peacefully in front of the mirror, staring for long periods of time. In two cases, subjects moved their hands in a circular pattern a few centimeters away from the mirror while watching both their hand and its reflection. In a few other cases, subjects looked into the mirror and then appeared to use it to look at their backsides.

We interpreted our observations as evidence that cotton-top tamarins have mirror self-recognition. In contrast to Gallup's mark test, our procedure caused a more salient change in the subject's appearance. This change may have caused an increase in the tamarins' attention to the mirror, a change that is necessary for recognizing the mirror image.[14]

Our results and interpretations were met with considerable criticism. Needless to say, if we were right, the great divide between great apes and monkeys would be in jeopardy. We therefore returned to the lab and ran a new group of tamarins on a modified version of the hair dye procedure, following a simpler design suggested in 1998 by the anthropologist Daniel Povinelli. The tamarins received less mirror exposure than in the earlier experiments, were tested in isolation rather than in their home room, and only one side of their hair was dyed. By dying only half of their hair we could, as in Gallup's original test, determine whether subjects would selectively touch the dyed half over the nondyed half. Under these conditions, we observed no evidence of hair touching while looking in the mirror, though we observed increased staring into the mirror in several individuals. Importantly, however, a control test elicited no self-exploration. We marked the arms of three individuals with bright green marks while they were anesthetized. When they awoke, they did not touch the marked area, though they looked at it. This failure suggests that the absence of touching in the original Gallup test may not reveal a lack of self-recognition. Rather, some species may lack interest in a novel mark. Though interesting, this should not be interpreted as evidence that tamarins lack self-recognition.

This most recent result reopens the debate. Perhaps, as Gallup, Povinelli, and others have argued, mirror self-recognition is restricted to humans and the great apes. Alternatively, perhaps the general failure in the second experiment is the result of individual variation among tamarins, variation that has also been noted for chimpanzees. It may also be due to subtle differences in experimental design. Either way, it brings us back to a point raised earlier. We must not be satisfied with single observations, either from a casual observer or a rigorous experimentalist. Replication is at the core of all science.

Studies using mirrors and television monitors for tests of self-recognition reveal a rift between humans, chimpanzees, bonobos, and orangutans on one side, and all other animals on the other. At present, we don't understand why some animals appear to recognize their mirror image whereas others do not. More important, we don't understand how a brain capable of self-recognition evolves into a brain capable of self-awareness.

A SENSE OF SELF

Most humans constantly struggle between the desire to maintain individual distinctiveness on the one hand, and on the other, to be part of a group and to be like its members. Individuality can be achieved in many ways, including the way we dress, speak, hit a ball during a tennis match, write, kiss, eat, laugh, and so on. Loss of individuality can also be achieved in a number of ways, including the adoption of a uniform dress code, haircut, religious belief system, corporate mantra, or a club's private language.

Certain forms of brain damage result in the loss of one's body image, a crucial component of individuality or personhood. As mentioned earlier, some prosopagnosics fail to recognize images of themselves. Capgras patients, in contrast, think that friends and close relatives have been invaded by foreign spirits. Among phantom limb victims, some think that the missing limb is present even though it is not, and quite often they feel excruciating pain in this area because they imagine the missing limb in a particular position. Here, then, the individual's body image is distorted. In a deficit called anosognosia, where one part of the body is completely paralyzed due to damage in the opposite hemisphere, patients are often delusional, arguing that there is no problem with the paralyzed side. For example, a patient with complete paralysis in one arm was asked to clap. Although only one arm engaged in this activity, the patient reported that she was clapping with both hands. As the neuroscientist Vilyanur Ramachandran muses in his book *Phantoms in the Brain*, this clearly answers the paradox of the sound of one hand clapping![15] Last, and as we shall discover in chapter 7, many autistics have virtually no access to their own beliefs or desires. They live in a socially isolated world, a glass bubble that blurs the distinction between self and other.

Recently there has been an apparent threat to our individuality, one that a large proportion of the public believes is unethical, a frightening sign that science has gone mad. The threat is mammalian cloning, particularly the creation of Dolly, a sheep cloned from an adult cell of one parent. Hollywood has, of course, always been interested in cloning, and has portrayed clones as evil, or at least as doing harm. Consider the clone in the classic movie *Metropolis*, who attempts to destroy a city, the dinosaur clones that run wild in *Jurassic Park*, the alien clones who start as seed pods in *Invasion of the Body Snatchers*, and the triple clone of Michael Keaton's character

Doug in *Multiplicity*, who each take on a different role in order to liberate Doug-the-husband. But Hollywood is Hollywood, and rarely concerns itself with getting the facts right. So what are the facts? What does cloning do to our sense of self, or, for that matter, any organism's sense of self?

When Dolly was cloned, she was, so the claim goes, a genetic duplicate of the parent cell. But here the shared sense of individuality stops. To make this clear, consider some real human clones: identical, monozygotic twins. We know, from the past twenty years of work by the psychologist Tom Bouchard, that when identical twins have been reared apart but are brought back together again, they show striking similarities, both in their material acquisitions and interests and in their personality traits:

> A pair we nicknamed the Giggle Twins laughed more than anyone else they knew, never engaged in controversial talk or voted, and had a habit of pushing up their noses with their fingers. Still another pair discovered they both used Vitalis hair tonic, Lucky Strike cigarettes, Vademecum toothpaste and Canoe shaving lotion. We have had two captains of volunteer fire departments . . . two women who would only enter the ocean backward and then just up to their knees.[16]

We are awed by these similarities because, of course, the twins were reared apart. So unless we are willing to grant them some form of telepathy, implemented at birth and throughout large portions of their adult life, then we must assign such similarities to their underlying genetic similarity. Putting our awe to one side, we must also remember that when these twins are tested on standard tests of personality and IQ, the correlation between their scores is about 50 percent; fraternal twins who were reared together show correlations of approximately 25 percent when given the same tests. What these results reveal is that the genetics account for a significant proportion of the variation, but not all. For identical twins, the remaining 50 percent of the variation must be due to environmental effects, either highly specified and controlled, or random factors. Returning to artificial clones, the message from twin studies is that no human clone could ever have the exact same sense of self as the parent cell. From the first sparks of fetal development, there are significant interactions between genome and environment, leading to an individual with its own unique mental signature, at least at some level.

Clones and identical twins have separate brains and bodies. What happens, though, when you have two brains but one body, as occurs naturally in Siamese twins? Is the sense of self more ambiguous, less differentiated? Does the answer to this question depend on how many body parts are unique or does all rest on having two heads, two minds? There are now many cases of Siamese twins, and all report differences between the twins. Several also report that each twin seems to have almost telepathic access to what the other is thinking. If true, it suggests a sharing of thoughts, and thus a sharing of self. And this logic should apply to animal clones and Siamese-equivalents.

No one has yet studied the psychology of Dolly, but one could imagine an experiment to see if a clone distinguishes its mirror image from its parent, or a picture of itself from a picture of the parent. Similarly, no one has studied the extent of psychological differences between identical and fraternal twins among animal populations, even though such cases are far more numerous than current mammalian clones. But there is a case that is analogous to the Siamese twin problem. It is a story about a two-headed black rat snake, named "IM" for "Instinct" and "Mind." It is a unique case, a sample size of one. Ordinarily we ought to be skeptical of single cases, but the two-headed snake opens the door to some fascinating psychological problems.

In the 1980s the comparative psychologist Gordon Burghardt studied IM, running various tests to determine the kinds of conflict it might experience during situations involving a choice. When IM was confronted with prey, the left head struck first for small prey, whereas the right head struck first for large prey. Regardless of prey size, however, both heads often attempted to eat at the same time. This move, however, cost the snake dearly. When they fed at the same time, ingestion lasted about an hour, whereas when they fed alone, ingestion lasted a few minutes. Given this cost, a savvy two-headed snake would develop manners, following a simple rule: Let the left head eat first when prey are small, but the right head when prey are large. In this case, two heads are not better than one. IM cannot access what it knows and believes; even more simply, it cannot learn from the many experiences it has had. Although it pays to have manners, gluttony and impatience win out. In essence, the right head is not only unaware of what the left head thinks, but also incapable of predicting what it will do based on experience. Here, then, is a potentially fascinating case, not because it provides us with evidence of self-awareness in animals—it patently fails on this count—but because it gets us to think about the right sort of questions.[17] Specifically, to examine the problem of self-awareness in animals we need to examine cases where an individual has private access to some kind of information and might use this information to model the world, assess the future, reflect on the past, and so forth. In essence, we need to find cases where the animal understands the reasons for its actions. and specifically, the causes of its desires, beliefs, and intentions.

All animals have a mental tool for recognizing others, distinguishing males from females, young from old, and kin from non-kin. Only a small number of animal species have evolved a self-recognition tool, one that enables them to distinguish self from all other entities in the world. Of this smaller subset of animals, our own species may be on its own in having the capacity to understand what it's like to have a sense of self, to have unique and personal mental states and emotional experiences.

All animals are equipped with mechanisms for learning, even though there are different schools of learning in the animal kingdom. Each classroom represents a different kind of learning, including imitation, teaching, trial and error, and deduction.

6

Schools of Learning

A female Japanese macaque drops a heap of wheat and sand into the ocean, and then skims the wheat off the surface once the sand has settled to the bottom. Although this technique is now a tradition in the population of monkeys living on the Japanese island of Koshima, it was invented by a highly creative female and then acquired by other members of the population. Naïve blue tits, watching skilled birds remove foil from a milk bottle and then drink the rich cream from the top, will then operate the foil in the same way. Human infants, only one hour old, stick out their tongues after an adult has performed the same display. An unmated female guppy will copy the mating preferences of another female if she watches the model's selection of males. These observations, all well documented, suggest that in group living animals, an individual's actions are highly influenced by social interactions. But how do these social interactions help in solving the problems of extracting food, choosing a mate, or finding safe refuge from predators?[1]

Assume you observed a Japanese macaque pick up a heap of wheat and sand, walk over to the water, drop the mixture in and then, as the sand drops down, skim the wheat off the surface and eat it. How did this monkey, or any of the others in the population, acquire the wheat-washing technique? One possibility is that a naïve individual walks over to the water without any group members in sight. Some wheat is floating in the water. She skims it off and eats it. All of a sudden she is struck by insight, and deduces the answer to this foraging problem. Like a contestant playing *Jeopardy*, she has been given the solution and must work out the question. The answer is, "The wheat floats, the sand sinks." The question is, "What happens when you bring wheat and sand over to the water and drop the mixture in?" In this scenario, deduction, not social learning, drives skill acquisition and thus knowledge. Here's a second, similarly asocial method of discovery: the animal walks over to the water for a drink and happens to have some sand and wheat on her hands. As she bends over, she watches as the sand falls to the bottom and the wheat floats. She acquires the skill through luck. Now consider a more social situation: a naïve individual watches as Imo, the actual inventor of this technique, picks wheat flecks from the water. The information is similar to our *Jeopardy* example, but there is a social factor. The naïve individual has once again observed the solution and must fill in the preceding steps. Now imagine a naïve individual sitting and watching as Imo picks up a heap of wheat and sand, heads to the ocean, drops the mix, eats the wheat, and then moves off into a tree. Several minutes later, the naïve individual reproduces this precise sequence of actions. Here, we would be tempted to conclude that the wheat-washing technique was acquired by imitation, a kind of recollection of prior events, a mental replay of the sequence. Finally, consider a case where Imo carries her infant over to the water's edge while the infant holds a mixture of wheat and sand in one hand. The mother then takes her infant's hand and places it in the water, allowing the infant to watch as the sand falls and the wheat floats. In this case, we might be tempted to say that the mother has taught her infant the wheat-washing skill.

When a population of individuals perform the same actions—like using water to separate wheat from sand—we must not jump to the conclusion that they acquired the skill, technique, or gesture in the same exact way. Although *all* animals are equipped with a basic tool for learning, there are differences in the design of this basic tool. For some animals, knowledge is acquired through deduction or trial-and-error learning. For others, it is acquired through imitation and teaching. The challenge is to understand which variants of the basic learning tool are available to which species, in which domains of knowledge, and why.

BABBLING BIRDS AND BABIES

Why don't birds communicate by means of English or Zulu? Why don't humans use song as their primary vehicle for communication? These may seem like trivial questions, but they are not. How does the nestling bird or human baby figure out which sounds in the environment it should attend to and which it should ignore so that it may build a repertoire of meaningful sounds? Do birds sing rather than chatter away in Zulu simply because song, rather than human speech, is what they normally hear? If so, then why don't birds raised in the company of humans deliver Zulu prose? Young birds and babies lack tutors, prepared to hold their hands and tell them what to listen for and when. And even if the pupils had tutors, adults wouldn't know what to tell them. When a mature bird sings and an adult human speaks, the information and rules that underlie these acoustic signals are generally inaccessible, tucked away in a part of the brain that is difficult to access, no less retrieve and explain to a naïve individual. Both bird and baby therefore travel the journey to communicative competence with little explicit tutoring. As with the other systems of knowledge that we have discussed, both bird and baby are equipped at birth with a general set of principles that cause them to attend to the sounds that are relevant to their species' repertoire. As long as the young individual is reared in an appropriate environment, it will acquire the native tongue. The question is, of course, what do we mean by an appropriate environment and what is the "native tongue?"[2]

To understand the similarities and differences between birdsong and human speech, we can borrow a trick from engineering. The trick is not, however, to build the device from scratch. Rather, we need to reverse-engineer the problem, working out what each system was designed to solve and the kinds of learning mechanisms provided to achieve the targeted goals.[3] Starting from a functional stance, one aimed at identifying what something is for, we know that all songbirds sing to attract mates and defend resources. In most species, only males take on this job. Hearing the song of a male from their species often causes females to approach and solicit further courtship, whereas males approach with the intent to attack and defend their turf. Song therefore conveys information about the bird's identity, motivational state, and in some cases, genetic quality. In contrast, although humans, both male and female, use speech to attract mates and defend resources, they also use it to gossip and pass down stories from one generation to the next.

How does the young nestling bird acquire song and the human infant acquire spoken language? Both animals hear a variety of sounds, but only some of it is meaningful. Furthermore, even the meaningful sounds are not all part of the native tongue—for example, laughter and sobbing are clearly important sounds for humans, but they are not part of the linguistic system. How, then, do young birds and infants solve these acoustical problems? An important part of the answer is that both species are guided by innate mechanisms —instincts—that constrain what they *can* learn. By "constrain," I simply mean limit the kinds of experiences attended to, as well as limit the range of possible outcomes. Thus, the nestling never learns to speak fluent Italian, and the infant never learns to sing like a lyrebird. However, nestlings do acquire a repertoire of song variants, and some humans acquire a repertoire of languages. In the words of the ethologist Peter Marler and the cognitive scientist Steven Pinker, all animals, including humans, have an *instinct to learn*.[4] Songbirds have a song instinct; humans have a language instinct. Instincts make complicated communication systems learnable.

Songbirds and humans are vocal throughout development. Along the path from birth to the first year, both birds and babies exhibit a universal pattern of development. They must hear the right kinds of sounds during a critical window of time, and then must practice what they hear. Both animals start with a babbling stage in which individuals produce sounds that approximate the adult form. Birds then terminate the process with a crystallized song repertoire, whereas humans terminate with a rich vocabulary and a set of grammatical rules for recombining words into sentences.

The mechanisms underlying song and language acquisition result in dialects, a learned style of singing or talking. Dialects often function as name tags. They typically identify where you come from and the group you are affiliated with. In the same way that human dialects can be lost, imitated, or molded by a tutor, the same is true of songbird dialects, at least for some species. At birth, there are no preset biases for picking up one dialect over another. Rather, the acquisition of a dialect depends on early exposure, particularly the kinds of sounds heard before the critical or sensitive period of development. Eliza Doolittle, the flower girl in *My Fair Lady*, was born with a cockney twang. If she had been raised by her tutor Henry Higgins, she would have an accent suitable for the BBC. Similarly, a white-crowned sparrow born in Berkeley, California, will sing a Berkeley dialect, whereas the same individual transported during the nestling phase to Tioga Pass, California, will learn the Tioga dialect. We must keep in mind, however, that in contrast to humans, there are nine thousand bird species, and song learning arises in only three of the twenty-seven major avian groups—parrots, hummingbirds, and oscines. Within this elite class of vocal learners, there are species differences in style of singing and in the details of the learning process. For some, such as the white-crowned sparrow and zebra finch, only one song dialect is learned during early development and then precisely reproduced during each mating season. Others, such as canaries and warblers, create new song variants each season in much the same way that Wagner created thematic variations or leitmotivs during such operatic masterpieces as *The Ring*. In both single- and multiple-dialect species, different populations maintain long-lasting song traditions, themes passed down from generation to generation.[5] The final class of song learners consists of the great mimics, species such as the mockingbirds, lyrebirds, and starlings. These species build an impressive repertoire of sounds, including songs from the local fauna as well as sounds from some of the inanimate objects in the vicinity. In the London area, a chaffinch learned to reproduce the ring of the British telephone company, and then appeared to use it as a prank to cause the master of the house to rush inside.

The computations that the avian brain carries out to acquire and then produce song are functionally similar to those carried out by the human brain in the service of language acquisition and production. Both communicative systems utilize dedicated brain areas. Both systems start out with circuitry that coordinates vocal production and perception, allowing the output from the mouth to be fine-tuned by the ear. For birds, the primary structure involved in production is the syrinx, whereas humans generate sound at the larynx and then filter it through the cavities above, terminating at the nose and mouth. During development, the bird and human brains undergo significant changes, showing the greatest level of plasticity prior to the onset of puberty or reproductive maturation. However, as neuroscientists such as Fernando Nottebohm have discovered over the past twenty years, some areas of the brain remain plastic even into the adult years. Thus, for example, as the canary changes song from one season to the next, key areas of the brain wax and wane in size as neurons are born and then die, perhaps correlated with forgetting old songs and learning new ones.[6] Although comparable findings have yet to emerge for human language, there is evidence for neural plasticity and reorganization in other domains, especially those having to do with body movements and sensations. In 1993 the neuroscientist Vilyanur Ramachandran showed that adult patients with a missing arm and the sensation of a phantom limb experienced a rewiring of the brain. For example, when the patient's chin was touched, he felt sensation in his chin and the "missing" or phantom pinky.[7] If the somatosensory area can be reorganized in an adult brain as a result of a phantom limb, then it is possible that some areas involved in language processing might be similarly reorganized following brain damage.

Let's return to our initial questions: Why don't birds acquire a human language, and why don't babies begin life singing like birds or, for that matter, speaking English or Zulu like adult humans? These questions demand answers at different levels, answers that I have provided in considerable detail in my book *The Evolution of Communication.* At one level, the mechanistic one, we must look at constraints imposed by the brain, vocal apparatus, and hearing system, both early in development and later on in life. What sounds can birds and babies generate and what sounds can they hear? At a different level, we must explore the adaptive function of the communication system, what it was designed for. Do good singers and good talkers reap reproductive benefits when contrasted with less articulate vocalists? At a mechanistic level, most birds can't produce human language because their vocal apparatus is not suited to producing the requisite sounds; even when a parrot articulates, "Hello, Polly wants a cracker?" he isn't generating the sounds in the same way that we are when we say the same sentence. As the psycholinguist Philip Lieberman pointed out in the late 1960s, the human infant is similarly limited in sound production due to the fact that she must use a vocal tract developed for digestive rather than communicative efficiency. At approximately four months, the human vocal tract undergoes quite significant changes, with the larynx descending deeper in the chest. As a result, the child turns into a talking machine, but becomes vulnerable to choking. In contrast with the avian vocal system, the human vocal tract can produce only the simplest, whistled bird songs, dropping out of the competition when it comes to producing different notes simultaneously, or trills that would turn our tongues into knots.[8] All of the above refers to mechanical constraints on the production of song and speech. What, however, are the psychological constraints?

Bird and human brains have specialized circuitry for processing and reproducing songs and speech, respectively. Damage to either the acquisition or production system can cause significant deficits in vocal performance. For example, if a bird is deafened after it has learned a dialect, its song will gradually deteriorate; if it can't hear what is being produced, the reproduction is extremely poor. But there are two critical differences between the acquisition process of birds and that of humans. First, while the capacity for imitation in birds is generally restricted to the vocal domain, it is more domain-neutral or indifferent in humans. Humans can imitate not only acoustic signals, but visual and tactile ones as well, winning the title of *Homo imitans* as well as *Homo sapiens*. Songbirds have a highly specialized faculty for vocal imitation, but cannot extend this capacity into other domains; parrots may represent an important exception (see the section entitled "Behavioral Clones"). When a bird receives the requisite input during development, such information is stored as a sequence of notes and syllables, ones that will define the bird's identity as a member of a particular species, from a particular population. In contrast, imitation in humans is highly promiscuous, allowing us to copy facial gestures, speech dialects, arbitrary hand gestures, and novel sounds tapped out with the hand, tongue, or lips. Further, when humans imitate an action, they often infer the model's intentions. Humans perceive actions as having goals and being guided by the actor's intention to achieve those goals. Thus, when we imitate, we are copying not only the physical action, but also the intentions underlying those actions. This ability to imitate actions and read intentions emerges before the second year of life.[9]

The nestling-baby comparison shows, once again, that similarity between two species can be misleading. Birds and humans have evolved the capacity to imitate. Whereas a bird's capacity is generally limited to sound reproduction, the human capacity is far more general. To work out how a mental tool works, we must navigate cautiously from our starting point in behavior.

LEARNING WITH FRIENDS

In 1980 the zoologist Holly Dublin recorded an unusual event while studying elephants in Kenya's Masai Mara Game Reserve. One of her focal subjects, an extremely pregnant elephant, started moving at a remarkable pace; her daughters followed behind. After walking several miles outside her normal home range, she stopped at a tree and feasted on almost all its leaves. Her daughters watched. The next day, she gave birth. Stunned, Dublin collected a few of the remaining leaves from the tree and headed off to one of the local villages. She approached a group of women, showed them the leaves, and asked if they were familiar. They giggled and informed Dublin that women use the leaves to induce labor during difficult pregnancies.

Dublin's observation represents a case of medicinal plant use. It is not an isolated case. There are now several examples in chimpanzees, woolly spider monkeys, and capuchins, earning this young field of research the formidable name of *zoopharmacognosy*. In terms of learning and the acquisition of knowledge, these observations are puzzling. How are medicinal practices passed on through the generations? Consider the elephant case again. Presumably, not all mothers have difficulty giving birth. Consequently, not all daughters would witness such a spectacular trek. But even if all mothers experienced difficulty, how would young animals learn what to do? Can mothers instruct their daughters if they see them struggling during pregnancy? Can they walk them over to the requisite tree and give them a helping of leaves? If young witness their mothers doing something as bizarre as walking outside the home range area, consuming all the leaves on a tree that they have never seen, and then giving birth the next day, can they figure out the cause-effect relationship—that eating the leaves helps with delivery? We have no answers for these particular puzzles, though there are others for which we do.

Much of the area now called Israel was once desert, but thanks to revolutionary agricultural techniques within this century, forests have blossomed, some filled with Jerusalem pine. This habitat has been invaded by black rats with a taste for the seeds from these trees. The seeds are, however, tucked away beneath the scales of the cones and the rats need a technique to extract them. The first clue that a technique had been invented emerged from observations of completely eaten cones lying beneath the trees. In the 1980s the biologist Joseph Terkel began a series of experiments to determine how black rats extract seeds from the pinecones and how their young acquire this skill. He placed untouched cones in cages with solitary adult black rats and watched. With exceptional skill, the rats peeled back row after row, extracting and eating seeds along the way. In contrast, naïve adults with no experience of pinecone stripping, or adults with experience stripping cypress rather than Jerusalem pinecones, never extracted seeds, or else they butchered the cone in the process of removing a few seeds. These observations suggest that black rats require experience to strip Jerusalem pinecones.

Naïve pups reared in the presence of mothers stripping cones learned the stripping technique. This shows that social interactions with a skilled mother have an effect on the pup's ability to strip cones. But is such exposure necessary? Not quite. Naïve pups and adults can learn the stripping technique if a human experimenter provides them with a mini-tutorial, starting with the presentation of cones where most of the rows have been removed, progressing to cones with only a few rows peeled away. Naïve individuals learn to extract seeds by working out the significance of cones with stripped rows, deducing the technique without any social input.

This population of black rats relies on pinecone stripping for their survival. Although naïve adults can deduce how to strip pinecones after being exposed to ones with stripped rows, this situation rarely arises in nature. In contrast, pups frequently confront this situation while watching their mothers eat, and may even have the opportunity to steal partially stripped cones from her. Terkel's field and laboratory studies therefore indicate that rats can learn pinecone stripping from either social or nonsocial factors. Pups reared by mothers who strip cones learn the skill, whereas those reared by naïve mothers do not. However, exposure to partially stripped cones in the absence of other rats provides sufficient information for naïve pups to deduce the technique.[10]

Pet rocks, phat baggy pants, *The Simpsons*, world music, hula hoops, happy face icons, the Beatles, Smurfs, punk, grunge, rap, burger chains, body piercing, cellular phones, baseball caps worn backward, Beanie Babies, deconstructionists, liberals, Double-Bubble, hippies, micro-breweries, Jordan-esque baldness, blue eyeliner, Kate Moss look-alikes, and Starbucks. Fads are pervasive. They come and go in a heartbeat. They are socially constructed, but tap into a deeply rooted biological disposition to be part of a group. Just when we have adopted one fad, another comes along and usurps it. Although animals may not be as readily susceptible to such whims, their preferences are often influenced by what others choose. These effects are most powerful in the context of food and sex.

Foraging decisions in the common laboratory rat are guided by an intricate system of information transfer. When a pup lacking specific food preferences is provided with a choice of chocolate- or cinnamon-flavored food, it picks chocolate if its mother did. The pup maintains this preference even if it has never seen its mother feeding, and even if the diet is good for the mother but bad for the pup. The pup's preferences are initially determined by chemicals transmitted from the mother's bloodstream, and then through the mother's milk. The pup also receives relevant information from the mother's breath and from the food scraps transported on the whiskers of other foragers. In a series of elegant experiments, spanning over twenty years, the comparative psychologist Jeff Galef and his colleagues have revealed that a naïve individual's food preferences are influenced by familiar as well as unfamiliar rats, an anesthetized rat that is unconscious, and a rat that has been made sick following food consumption. Further, the food choices of a relatively sick animal are more significantly influenced by a demonstrator than a healthy animal, and naïve rats are more likely to follow a leader through a maze if the leader has fed on a good safe food than on a poisonous one. These examples show that rats are social eaters, choosing foods that have been road tested by others. Unfortunately, such social effects often lead to deleterious consequences, as when a rat chooses to eat what it finds on the whiskers of a dead rat. The rat is equipped with a learning mechanism for food that generates intelligent errors—something we also found in our discussion of spatial knowledge (chapter 4). Under normal circumstances, a pup does well by eating what its mother eats. Faced with the devious manipulations of an experimental psychologist, however, this intelligent mechanism looks remarkably stupid.

For the common rat, then, social interactions provide a cost-free forum for sampling available foods in the environment. Each interaction provides information about potential foods, items that fellow rats have eaten and therefore survived. By tapping into this network of information, naïve rats build a repertoire of food items without trial-and-error sampling.[11]

The cases of social learning that we have examined thus far focus on the context of feeding, and on two species of rodents. Such social learning is far more ubiquitous, appearing in octopi, bluehead wrasse, red-winged blackbirds, baboons, and humans.[12] There are also other contexts, especially when following the lead of an experienced individual pays off. Choosing a mate is perhaps the most obvious context. In a wide variety of species, females are finicky, spending a great deal of time looking for and then selecting an appropriate mate. Prior to making a selection, females assess their options, using anatomical and behavioral traits to identify the best possible mate. Thus, we know that females tend to mate preferentially with large over small males, with symmetrical over asymmetrical males, and with males producing low-pitched vocalizations over those producing high-pitched vocalizations. These traits provide useful information about male quality, information that females use in picking a mate. Because it takes time and energy to find a good mate, natural selection will favor mechanisms to reduce such costs. One possibility is that some females copy the mating preferences of other females.[13]

Guppies spend their lives in large schools. From studies carried out in the streams of Trinidad, as well as in laboratory choice tests where a female is given the opportunity to approach and mate with one of two males, we know that females prefer to mate with brightly colored males over dull colored ones, choose mates with low parasite loads over ones with high parasite loads, and prefer symmetrically spotted males over asymmetrically spotted ones. In 1992 the biologist Lee Dugatkin set up an experiment to determine whether a female's choice of mates might be influenced by the choice of another female—what has been called *mate choice copying*. An observer female watches as a model female chooses between two males housed in separate tanks. Next, the observer is presented with the same choice. Consistently, the observer selects the same male as the model. This effect, however, depends critically on the presence of courting males. If the males are replaced with two females, or with males that do not court the model female, the observer shows no mate choice preference.

Like naïve rat pups, naïve female guppies are significantly influenced by the mating preferences of a model. But how vulnerable is a guppy to social pressures if she has already selected a mate? Dugatkin ran a second experiment involving three steps. In step 1, a female observer Jane chooses between males Sam and Bob. If she chooses Sam, she then watches as a model female Ellen picks Bob in step 2. In step 3 Jane has to again choose between Sam and Bob. Under these conditions Jane reverses her preference, picking the same male that Ellen selected—Bob. Interestingly, observers are more likely to reverse their choice of mates if the model is an older female than if she is a younger female. Alas, age and wisdom are coupled in the world of guppies.[14]

The work on rats and guppies brings us back to a point raised at the start of this chapter, one that is equally applicable to birds, dolphins, rodents, monkeys, and apes. There are a number of ways for one animal to act like another animal. A black rat may acquire the pinecone-stripping technique by observing its mother or by receiving an experimentally orchestrated tutorial. A rat pup may acquire a preference for chocolate- over cinnamon-flavored food because of a chemically transmitted message from her mother's milk, from following a stranger to the chocolate-flavored food dish, or from smelling a sibling's breath. Guppies may prefer brightly colored males over dull ones because of a genetic disposition toward such choices or because they witnessed an older and wiser female making this choice in front of them.

Important questions remain. When animals copy a skill, such as wheat washing in Japanese macaques, do they store a mental record of the demonstrator's actions? If they mentally represent the action, do they also encode the actor's goals, his intentions and putative desires?[15] Do animals imitate?

BEHAVIORAL CLONES

The rain forests of central Kalimantan, Indonesia, are home to a large arboreal ape, the orangutan. Two populations of orangutans are particularly well studied, thanks to the efforts of the anthropologist Birute Galdikas and the psychologist Anne Russon. One population is wild, its only contact with humans coming from occasional encounters with local Indonesians walking in the forest and researchers who have come to study their behavior. The other population, rehabilitants, consists of orangutans reared by humans and then reintroduced to the forest with the hopes that they will start life over again, feeding on the local fruits and restocking their gene pool. As one might imagine, reintroduction is a slow process. After all, these animals have often been coddled, reared by caring humans. As a result, although they cannot forage on their own and do what orangutans have been selected to do, they have picked up some peculiarly human characteristics. In particular, in the early 1990s Russon and Galdikas observed several cases of what they call *spontaneous imitation*, cases where one individual appears to reproduce an exact replica of a previously observed human action, including the appropriate use of a toothbrush as well as insect repellent.[16] Consider the observation that an orangutan named Supinah acquired the ability to ignite a fire with kerosene and then tend to it by waving a trash can lid over the burning embers. No one taught Supinah these actions, and no one rewarded her for them either. How then did she acquire such fire-making skills? Russon and Galdikas suggest that the orangutans learn by imitation, by performing an act from seeing it done. This is a potential interpretation of Supinah's actions, and a potential definition of imitation. For some researchers, the imitated act must be novel, a behavior that is unlikely to be performed as part of the daily routine. For others, imitation depends on a capacity to read another's intentions, to work out the goals and plans of the demonstrator based on their behavior. Returning to Supinah, we know that fire construction is a novel act, as no orangutan has ever tried such a stunt in the past. Nonetheless, we can ask about other aspects of the reproduction, especially the fidelity of each step in relation to the presumed demonstrator, the campsite's cook. For example, did she copy an entire sequence, or did she copy components of the sequence and then string them together in a novel fashion? Did she attempt to emulate the cook's goal of lighting a fire, or the steps involved in achieving the goal—pouring kerosene and igniting a burning ember with kerosene? Does she even understand the goal, or does she merely copy the steps, only to be surprised when her burning stick ignites into flames? Although Russon and Galdikas's observations suggest that the rehabilitated orangutans imitated both the form of each action and the order in which each step of the action is carried out, such claims are difficult to make in the absence of more detailed case histories. What may appear to be the first instance of spontaneous imitation to a human observer may in fact be preceded by a long run of trial-and-error sampling.

How sophisticated is a human newborn? For starters, it can barely see, can't locomote on its own, can't speak a language, doesn't recognize itself in the mirror, doesn't understand witty repartee, and is emotionally volatile. Quite unsophisticated. But within less than an hour after their umbilical cord has been cut, babies begin imitating facial gestures and finger movements. How can this be?

Over the past twenty years the psychologists Andrew Meltzoff and Michael Moore have been chiseling away at the problem of human imitation, attempting to show how this capacity develops. The catalyst for their research was the discovery that when newborns less than an hour old watched adults performing a display—sticking out their tongue, opening their mouth in an O-shape, protruding their lips, or flicking their fingers—they matched the particular display after the adult stopped.

Imitation of facial displays is different from imitation of finger flicks or Supinah's fire preparations. All forms of imitation involve seeing an action and then repeating it later. In the case of facial expressions, however, the imitator lacks visual access to his performance. He can feel what he is doing, but cannot perfect the action based on visual feedback. He needs a mirror, but one inside his head, accessible to his mind's eye. Facial imitation therefore seems to require the capacity to record a script of the demonstrator's expression and then access this record during the imitative reproduction. But if the actions are part of the neonate's natural repertoire, things it does on its own without prompting, then perhaps seeing a display primes the infant to perform one of these actions. Imitating an unusual or atypical action would be more impressive, more convincing.

Meltzoff and Moore showed six-week-old infants an atypical facial display: an adult sticking his tongue out and to the side of the mouth. Infants spontaneously responded by protruding their tongue to the side, but often failed to produce a precise copy. Over time, however, the infants' reproductions improved. This suggests that infants maintain a representation of the target response, and attempt to match it during their own reproductions.

Given that the infants were not rewarded with a food pellet for their tongue protrusions, why were they doing it? Infants don't know what facial expressions mean, what they have been designed to express. As the psychologist Paul Ekman suggests, however, voluntarily moving one's face into a particular expression elicits the corresponding emotion—happiness with a smile, sadness with a frown. Given these findings, Meltzoff and Moore suggest that when babies imitate, they are unconsciously tapping into this association. They are also tapping into something much more important. Facial imitation provides a mental tool for communicating with others, especially when spoken language is not an option.

To test their explanation about early infant imitation, Meltzoff and Moore ran the following experiment. An infant sat and watched as two adults presented, one after the other, different facial gestures. For example, Fred walked in and protruded his tongue to the side. Joe then walked in and protruded his lips. During these displays, the infant sucked on a pacifier, which prevented immediate imitation. Twenty-four hours later, Joe walked in and stood, expressionless, in front of the infant. The infant protruded its lips. Then Fred walked in and stood expressionless. The infant stuck its tongue out and to the side of its mouth. It's as if the infant was saying to Joe "Are you the lip-protruding guy?" Imitation allows the infant to establish individual identity, confirming whether the person standing in view is the same person encountered the other day.

By the age of fourteen months, infants will imitate actions that they saw performed four months earlier, even when they are extremely bizarre ones such as an actor bending down at the waist to bang a panel with his head. Infants at this age also notice when others are imitating them. Infants smile more, look more, and direct more of their responses to an actor who is imitating their every action than an actor who moves when the infant moves, but does something else or does the same thing at a different time. Infants know when someone else copies the precise form of their own actions.

Given that imitation provides a tool for identifying individuals, might it also provide a tool for identifying an individual's beliefs, desires, and intentions, mental states that can be revealed through actions? Meltzoff set up an experiment with eighteen-month-old infants. In one group, infants watched as an adult successfully acted on an object. For example, they reached for a jar and successfully removed the lid. The jar, with its lid secured, was then handed to the infants. Infants readily imitated the actor's specific actions, removing the lid from the jar. In the second group, infants watched as an adult tried but failed to achieve the intended act. For example, they reached for a jar and failed to remove the lid because their hand slipped off the edge. Here, although infants failed to witness the successful act, they nonetheless reproduced the demonstrator's action, extending it to its successful end point. Infants imitate actions even when they fail to see a successful demonstration. They appear to read the actor's intentions, recognizing that actions are guided by desires and goals. But is this conclusion warranted? Perhaps infants spontaneously figure out the target action without watching the demonstrator? If so, then we might not want to conclude that they are imitating the actor's intentions or desires. After all, we know that young infants distinguish animate from inanimate objects and generate different expectations about them (see chapter 7).

Meltzoff ran two key controls to test these ideas. In one group, infants watched as an adult manipulated an object, but without a successful or an unsuccessful action on or toward the object. For example, they rubbed the back of their hand against a jar. These infants failed to show any goal-directed, target actions. A second group watched a robot fail to perform a successful action toward an object in the same set of circumstances as the human actor. For example, the robot failed to remove the lid from the jar. As in the first control, infants failed to show goal-directed, target actions.

Significantly, then, infants read the intentions of a human actor, using these to guide their own imitative actions. Inferring the actor's intentions and goals is part of the imitative process for human infants.

Studies of wild birds, dolphins, monkeys, and apes suggest that imitation is prevalent in the animal kingdom. Such observations are, however, either fraught with interpretative problems or difficult to reproduce under more controlled conditions. In the wild, researchers can only rarely document with confidence how a skill or gesture was acquired; more typically, scientists encounter an animal with a skill in place rather than an animal in the process of acquiring one.

Consider a classic in the field of ethology, the observation that blue tits in the London area learned to remove aluminum foil from the 1940s-style milk bottles, and then skim the cream from the surface. Given the rapidity with which this skill appeared in the population, the ethologists John Fisher and Robert Hinde concluded that naïve individuals imitated the technique from accomplished demonstrators. In 1984 the psychologists Jeff Galef and David Sherry trained a group of black-capped chickadees to remove foil from a milk bottle and then drink the cream. One group of naïve chickadees watched the experts remove foil from the milk bottles. A second group of naïve chickadees received an open bottle of milk and a piece of foil sitting next to it; this allowed them to skim the cream from the surface on the first trial, without having to remove the foil. On the second trial, they received a closed milk bottle. Although individuals in the second group participated in experiments without a demonstrator, they acquired the foil-removing maneuver as rapidly as individuals in the first group. It is possible, then, that the London blue tits learned their skill in the absence of social input, using straightforward deduction from a set of helpful cues.[17]

When a naïve bird learns to remove the foil from a milk bottle after watching a demonstrator, can we conclude that the naïve individual imitated the demonstrator's actions? If so, what components of the action did it imitate? Unlike the experiments by Meltzoff and Moore with human infants, the chickadee experiments lack crucial information; thus, we cannot make any conclusive statements. For example, since we don't know precisely how the demonstrator removed the foil, and whether there are different ways to do it, we don't know if the observers were replicating all the requisite steps or style, whether they were simply attracted to the foil and then worked out how to take it off on their own, or whether they inferred the demonstrator's intentions from watching. Would naïve observers do as well with a human demonstrator or a robot? More detailed experiments on rats, parrots, and chimpanzees hold some of the answers.

In the 1950s a chimpanzee named Viki was raised in a house by humans and was taught to play an imitation game. In the early phase of this work, Viki was trained to repeat the human trainer's action whenever she heard "Do this"; when Viki had trouble, the trainers would help by shaping her hands or body into the appropriate position. Once she learned this game, novel actions were presented, and Viki copied them. When the task involved actions on an object, actions that were designed to solve problems, she generally failed. Since the Viki studies, several other researchers have tried the "do-as-I-do" technique with captive chimpanzees and orangutans. Whereas both species of ape can imitate arbitrary body movements, they fail to imitate actions that have a goal-directed component such as opening a jar.

Other attempts to show imitation in primates have met with little success, especially tests of capuchin monkeys and chimpanzees. In one experiment by the psychologist Mike Tomasello, a human demonstrator grabbed and then used a T-shaped rake to access small pieces of food placed out of reach. Chimpanzees and two-year-old children watched. For each species, one group watched a demonstrator using an inefficient technique, whereas the other group watched a demonstrator using an efficient technique. Following the demonstration, subjects were given a rake, with food placed out of immediate reach. The goal of the experiment was to determine whether chimpanzees and children would imitate the demonstrator's actions or emulate the demonstrator's goals.

The chimpanzees and children both learned to use the rake to access food, but there were striking differences in their techniques. The children used the precise technique demonstrated, whereas the chimpanzees used whichever technique worked for them. The children's *loyalty* to the demonstrator's technique is surprising given the fact that for some, the raking action was inefficient. Here, then, chimpanzees learned something about the rake's connection to food, but unlike the children, they failed to imitate the steps in between grabbing the rake and eating the food.[18] In comparable tests run on monkeys, social interaction helps individuals reach the target goal, but they don't imitate the demonstrator. Further, there is no evidence that monkeys or apes imitate nonvocal gestures, and only suggestive evidence that they imitate vocalizations. At present, we lack a satisfactory explanation for this hole in the primates' toolkit, for why their capacity for imitation is so severely compromised.[19]

Recent work on a gray parrot serves to remind us that we should look for parallel capacities in species that confront similar environmental pressures. Parrots and primates represent a case in point. Parrots live in large social groups, and like many of the great apes, have also been reared in close contact with humans.

Like many parrot owners, the psychologist Bruce Moore developed a close bond with his parrot and provided him with an environment that readily allowed for human speech to be imitated. Surprisingly, however, the parrot learned to spontaneously imitate not only a number of phrases, but also the gestures that accompanied them. Thus, the bird learned

> to say "ciao" and to wave good-bye with its feet or its wings. . . . When it refused a nut, the experimenter sometimes tossed the offering aside. That too was imitated, though never with an actual peanut; the bird simply reproduced the movements while repeating the labeling phrase, "Forget it!" Whenever the bird dropped a nut, the experimenter bent down to retrieve it. The bird later copied that while repeating "Whoops, dropped the peanut." . . . And the bird learned to say, "Remember Lloyd Morgan, don't forget," with the last two words accompanied by emphatic finger (claw) strokes.

What is most impressive about the parrot's imitations is that they were spontaneously generated. Unlike the do-as-I-do work with the apes, which involved an initial training phase, Moore's gray parrot picked up the demonstrator's speech and gestures, and not only any gesture, but the ones most appropriate to the utterance produced. Moore concludes that his gray parrot imitates in both the auditory and visual modalities. As it is for human infants, imitation is fueled by a clear social payoff. Humans find the parrot's parroting delightful, even magical. As social creatures, parrots reap the benefits of our delight, as we spend hours and hours in pet stores trying to get poor old Polly to say anything.[20]

A crucial feature of Tomasello's rake experiment is that the goal (eating food) is distinguished from the process of obtaining it (raking), and the process can be achieved by at least two alternative techniques—efficient and inefficient. Consequently, we can distinguish between individuals learning the goal and individuals learning the steps required to achieve the goal. In contrast, the do-as-I-do procedure used with chimpanzees and Moore's gray parrot provides only one target gesture, and only one way to produce it. It is therefore difficult to say whether these two species emulated the goal—a gesture—or the steps involved in producing the gesture.

Using the two-choice technique, the psychologist Celia Heyes has demonstrated that rats will use whichever technique an actor demonstrates. In the rat experiments, if the action of moving a lever to the right or left occurs without a demonstrator—ghostly action by means of experimental magic—then the observer fails to copy, although it does learn the task. Somewhat like Meltzoff and Moore's infants, rats appear to require at least some kind of demonstrator performing the action, although it is not clear whether they would tolerate an animal other than a rat, or even a robotic rat.[21] This kind of experiment is critical for our understanding of how animals represent the action observed. Are animals reading the actor's intentions? Are they simply attracted to the device that is being acted on, rather than the actions performed by the demonstrator? If the subject copies an action in the absence of a demonstrator, then it has certainly learned something from the demonstration, but we no longer need to talk about imitation.

A twist to the imitation story emerged in the early 1990s when Mike Tomasello, Andrew Whiten, and their colleagues initiated a suite of studies using both normally reared apes and apes reared by humans. The motivation for this comparison was to determine whether the human environment might, in some way, enhance the chimpanzee's capacity for imitation. Perhaps chimpanzees have the capacity to imitate, but lack the relevant environmental experiences to trigger this capacity.

In a series of experiments, Tomasello and his colleagues showed that human-reared chimpanzees spontaneously imitated a human's action on an object in about 40 percent of all presentations; comparable experiments on one-and-a-half to two-and-a-half-year-old children showed slightly higher levels of imitation, whereas naturally reared chimpanzees failed to imitate the human's actions. These results led Tomasello to the conclusion that *enculturation*—the process of rearing animals in a human environment— enriches the ape mind, elevating the capacity for imitation to new heights.

Opposing the enculturation hypothesis are studies of naturally reared chimpanzees by the psychologist Andrew Whiten. Whiten created the equivalent of a Rube Goldberg box, what he called an "artificial fruit," consisting of a Plexiglas box with a food reward secured inside. To access the food, naturally reared chimpanzees had to manipulate several bolts, pins, and handles. Each component offered two optional techniques, such as twisting or poking the bolt. The chimpanzees watched three box-opening demonstrations by a human experimenter, and then had an opportunity of their own. Three of the four chimpanzees used the technique demonstrated, though not all components of the opening technique were copied with the same fidelity. Whiten considers these observations evidence that chimpanzees imitate, even in the absence of enculturation.

Do animals imitate or don't they? Is enculturation necessary? These questions are still live, and much more work lies ahead. Much of the debate concerns a technical distinction between goal emulation and imitation, whether animals copy the end product of an action or whether they copy the precise action used. If they copy actions, why do some actions appear more difficult to reproduce with high fidelity? What seems clear is that monkeys don't imitate, rats and parrots may, and some chimpanzees do, at least under certain conditions. No animal, however, imitates the way a human does, even a very young one. The imitation tool is a powerful one to have in one's mental toolkit. It makes stupid individuals look smart, and provides one of two efficient tools for creating cultural traditions.[22] The second tool is teaching.

TEACHER'S PETS

A child reaches for the bright red glow of a stove-top burner; the mother shouts, "No!" and pulls the child's hand away. A tennis pro stands behind his pupil as they both hold onto the same racket and hit a stroke in unison. A university professor lectures to an audience of students taking down notes. A grandmother allows her granddaughter to continue a knitting project she has started.

In each of these cases, an individual with considerable knowledge and expertise conveys their wisdom to an ignorant individual, teaching them. By using such an umbrella term as *teaching*, however, we risk covering up important differences in the kinds of thoughts and motives underlying each action. Thus, for example, a mother may slap her child's hand as a reflex, an unconscious, automatic action that pays off precisely because it doesn't involve planning or forethought. Imagine what would happen if the mother, upon seeing her child reach for the stove, thought, "Hmmm, my daughter must be putting her hand next to the burning hot stove because she doesn't understand the significance of the red flame. I had better teach her a lesson so that she doesn't hurt herself now and in the future. I will run over and slap her hand away." Although this would certainly work in some cases, the situation is urgent, demanding a swift and nearly automatic response. Although such responses lack explicit planning, they can be functionally instructive. In contrast, the tennis pro's actions are clearly pedagogical instructions based on the recognition of ignorance. By standing behind the pupil and guiding her hand, the instructor molds the stroke, providing the requisite information for hitting a good forehand. The instructor's actions are intentional, designed to achieve a particular goal. If animals teach, therefore, we need to assess whether they do so reflexively or with foresight and planning.[23]

Skilled use of a tool often requires several years of practice. One of the more striking examples of this among animals is the hammer and anvil technique employed by chimpanzees in western Africa and documented in the 1990s by the ethologist Christophe Boesch and the comparative psychologist Tetsuro Matsuzawa (see chapter 2). The process starts with finding a functional hammer and anvil. The hammer is either a branch or a stone, configured with a thin grasping end and a thicker striking end. The anvil is either a flat stone or a log, one that will hold a palm nut in place while the hammer comes down and cracks it open. Chimpanzees begin cracking open nuts with some success at around three and a half years, but do not acquire the requisite competence for several more years. Given how long it takes to learn the nut-cracking skill, one might imagine that such pressures would set up the necessary conditions for teaching to evolve, for skilled adults to tutor their young in the art of hammer and anvil operation.

When mothers gather palm nuts and set up at a nut-cracking site, their young often sit around, watching. But the young do more than watch. They often steal nuts from the anvil or from the mother's general stash. Sometimes mothers leave their hammer and anvil set in the presence of young, thereby providing them with an opportunity to manipulate the tools. We certainly recognize such opportunities as invaluable to learning, though they represent only a weak form of teaching. Stronger, but extremely rare,

are two cases where chimpanzee mothers have corrected their offspring's attempts at nut cracking. In one case, an infant picked up a branch, fat at one end and thin at the other. With her nut placed on the anvil, she held the fat end of the branch and attempted to crack the nut with the thin end. After the infant failed several times, her mother approached and reoriented the branch. On the next try, the infant held the thin end, struck with the fat end, and cracked the nut. In the second case, a mother helped her infant reposition the nut, placing it on a more secure part of the anvil. Her infant then successfully cracked the nut.

These observations suggest that chimpanzees have the *potential* to recognize inappropriate behavior in another and to respond in such a way that they can make corrections, behavioral fixes that presumably lead to greater competence. I say "presumably" because we actually don't know whether these two cases of instruction resulted in more efficient nut cracking over a longer period of time. So why are such cases of instruction rare, only two cases in more than 150 years of chimpanzee observations? Given the trivial costs incurred by the mother for this brief instruction, and the potentially significant benefits accrued from learning what to do, one would expect pedagogical interactions to be more frequent. Without knowing a great deal about the costs and benefits associated with chimpanzee instruction, we can only speculate here. But two additional factors may shed some light on the problem. Specifically, although young chimpanzees appear to learn from instruction, mortality rates are low during the juvenile years, and are rarely due to food shortages. Thus, although the technique requires years of experience to perfect, there may not be any pressure to learn the skill earlier in life. Also, although instruction helps with certain aspects of the nut-cracking skill, sufficient strength is also required, as well as an ability to find nuts and an appropriate hammer-anvil set. These motor and cognitive capacities develop slowly, and it is not clear how instruction would speed up the process. Before we puzzle over this problem any more, let us explore some additional terrain, attempting to identify other candidate examples.[24]

The Serengeti National Park in Tanzania is populated with thousands of wildebeest and gazelles, as well as elephants, giraffe, buffalo, baboons, and of course, the big cats, leopards, lions, and cheetah. In contrast to lions, who hunt cooperatively, leopards and cheetah tend to hunt alone as adults. If hunting is a solitary affair, how do individuals acquire the requisite skills? Is it by trial and error, a kind of hit-or-miss operation? Or do young learn from their mothers?

Tim Caro, a wildlife biologist, has been studying the Serengeti cheetah since the 1980s, following their moves from the safety of a Land Rover. To understand how hunting skills develop, he started watching cubs, mothers, and females without cubs. When females without cubs chase after a Thompson's gazelle, they kill it by suffocation; if they are after a hare, they bite it through the skull. Mothers with one-and-a-half-month-old cubs hunt in the same way. But when the cubs are about three months old, mothers hunt with a different style. They chase after the prey, first maiming it and then carrying it back to the cubs so that they have the opportunity to first chase and then knock it over. Because cubs at this age rarely kill the prey, mothers intervene, finishing off the job. When the cubs are a few months older, mothers release only one-third of their prey in the presence of their cubs and allow them to finish the kill. At around eight months, cubs rip the prey apart while another member of the family bites into the esophagus to keep the prey from moving off. By ten months, the mother releases half of her prey and cubs are now almost completely successful in the final kill.

What these observations show is that cheetah mothers provide cubs with an opportunity to acquire hunting skills in a controlled environment, what one might call *opportunity teaching*. Their behavior appears sensitive to the cubs' developing abilities; the kind of opportunity presented changes as a function of current hunting skills. But in contrast to the chimpanzee case, cheetah mothers incur a significantly greater cost. Prey released to cubs are much more likely to escape, thereby forcing the mother to go out on another hunt. Hungry mothers therefore are less likely to repeatedly release prey than relatively satiated mothers. By doing the hunting themselves, hungry mothers insure a catch, but eliminate the opportunity for learning. Although we don't know whether cubs with fewer opportunities to hunt end up being less successful hunters later in life, it appears that such opportunities must be beneficial, given the kinds of costs that mothers are willing to incur.

The comparison between cheetah and chimpanzees allows us to make some predictions about the frequency of teaching in a population. Think of the problem in economic terms, where the teacher incurs certain costs of instruction and obtains certain benefits from the pupil acquiring a skill or some knowledge. Similarly, we can think about how the pupil benefits from instruction and the costs incurred if instruction is withheld. For chimpanzees, the costs to the teacher are low and the benefits to the pupil are relatively low; although infants appear to benefit from instruction, they clearly do not depend on instruction for survival, and mortality tends to be quite low during the juvenile years. In contrast, the costs of instruction are high for cheetah mothers and the benefits to cubs appear high as well.

Generalizing from chimpanzees and cheetah, we can make some specific predictions. When the benefits to the pupil are high, teaching should be common even if the costs to the teacher are high. When the benefits to the pupil are low, teaching should be infrequent when the costs to the teacher are low, and either rare or absent when the costs to the teacher are high.[25]

In the late 1960s the zoologist Tom Struhsaker initiated a study of vervet monkeys living in Amboseli National Park, Kenya. Like many other primates, vervets face intense threats from predators. Amboseli is, however, particularly nasty since the vervets fall prey to a spectacular cadre of predators including cheetah, leopards, pythons, black mambas, green mambas, baboons, humans, martial eagles, and crowned hawk eagles. Twenty years of observation reveal that 70 percent of all infants fail to make it past their first birthday. The primary cause of mortality is predation.

Fortunately for vervets, the exceedingly strong pressure from predation has resulted in the evolution of a sophisticated system of alarm calls (see chapter 8). Vervets have evolved a unique alarm call for each predator type confronted, calls that trigger the appropriate evasive action. Thus, when leopards are nearby, the best response for a vervet is to run up into a tree. Upon hearing a leopard alarm call, vervets run up into trees. When an eagle is swooping and an eagle alarm call has been sounded, the best response is to run out of a tree and under a bush. And if a snake has been detected or a snake alarm call heard, the best response is to stand bipedally and take a good look as to its whereabouts and possibly its direction of movement. Adults know these moves. Infants do not, at least not until they are about a year old or older. How do they learn?

When a vervet gives an alarm call, others often call as well. Consequently, when an infant spots a predator and gives an alarm call, repeat calls by adults function as positive reinforcement, functional pats on the back for a job well done. Although adults may not be intending to teach their young, their alarm calls provide positive feedback for the infant's vocal behavior.

When infants under the age of one year were the first to call, follow-up alarm calls by adults occurred about half the time. Further, infants appeared to benefit from such experiences, as shown by their performance on subsequent encounters. Infants who received positive feedback from adults were more likely to produce an appropriate alarm on the next encounter than infants who either failed to receive such feedback or called to an inappropriate target. As in many other examples emerging from animals living under natural conditions, we cannot know if we have properly documented the infants' learning environment. Although thousands of hours were logged by the vervet research team, it was not possible to log all their experiences.

If we move away from these more detailed studies of teaching in animals, we are treated to an enticing catalog of onetime observations. For example, one vervet infant saw a mongoose (a nonpredator) and gave a leopard alarm call. Immediately she was picked up by her mother, who headed up into the trees. Upon seeing the mongoose, the mother turned to her offspring and slapped her, in what looked like punishment. Among birds, the juvenile yellow-eyed junco is perhaps one of the most inept foragers. In contrast to adults, juveniles pick up mealworms with an inappropriate orientation and, as a result, often drop them. Occasionally, adults approach juveniles and reorient the mealworm in their beak, thereby allowing them to eat their catch. Last, studies of human-reared chimpanzees reveal a few instances of sign language instruction. Thus, for example, while Washoe was anxiously waiting for food with her daughter Loulis, a chimpanzee raised without sign language instruction, she grabbed Loulis's hand and molded it into the sign for "food" repeatedly guiding her through the relevant actions.

Returning to our economic calculations, we find that many of our observations of teaching appear cheap from the perspective of time and energy invested by the teacher. In many of these cases, it would appear that the payoffs for the pupil are high. We are thus left to wonder why teaching is so rare in the animal kingdom, but so common among humans.

All animals enter the school of learning with a core tool for acquiring knowledge. This tool is sufficient for building associations, making deductions, and learning by trial and error. Many animals also learn through social interactions. Surprisingly, perhaps, humans may be the only species to have evolved the mental tools for imitation and teaching.

The world of deception includes falsifying information by crying wolf, and concealing information by remaining quiet.

7

Tools of Deceit

In the rain forests of Peru, large flocks comprising different bird species scour the forest floor and canopy for food. Given the competition between species living in the same area, it is surprising that such mixed flocks move and feed together. For the evolutionary biologist, however, such situations immediately raise intriguing problems concerning the payoffs of group life. What do members of each species gain as a function of joining a mixed flock? What costs do they incur? Does it ever pay to defect and join a pure, single-species flock?

In 1986 the biologist Charlie Munn set out to address these questions and en route noticed something unusual. After hours and hours of observation, he noticed that the flock was divided into food-finding species and alarm-calling species. Although division of labor policies are common in such species as honeybees and naked mole rats, this mixed-species flock operated in a similarly coordinated fashion, each species contributing its own unique talents, raising the benefits of membership above the costs of living in a single-species flock. The last twenty or so years in biology have taught us, however, that whenever individuals assume different roles in an apparently cooperative venture, there are opportunities for individuals to cheat, reaping the benefits without paying the costs. Such cheaters have emerged in the mixed-species flocks in Peru.

Flying insects are a primary source of food for members of the flock. From time to time, an individual from the food-finding species competes over an insect with an individual of the alarm-calling species, either the bluish-slate antshrike or the white-winged shrike tanager. During some of these clashes, the antshrike and tanager sound a false alarm, signaling impending danger from a predatory eagle, although there is no danger. Nonetheless, the competitor looks up following the call, allowing the antshrike or tanager to snatch the insect away. On the face of it, one might think that such routines—straight out of a Three Stooges comedy sketch—would work once, maybe twice, but then the deceived would figure out the dishonesty of the signal and learn to ignore it. Unfortunately, the food-finding species can't ignore the alarm call. They are, to put it simply, in an economic bind. Compared to the trivial benefit associated with capturing one insect, the cost of ignoring a potential predator is high. Although antshrikes and tanagers give honest alarm calls often enough to pay their membership dues to the flock, they get away with false alarm calls because it never pays the competitor to question their honesty. Natural selection has thus given antshrikes and tanagers the upper hand, at least during this evolutionary moment in time.[1]

The fact that antshrikes and tanagers produce deceptive alarm calls says nothing about what the caller thinks under these conditions. We don't know whether they simply work out *how* to deceive or whether they also know that they are deceiving someone else by manipulating their beliefs. Does, for example, the tanager consider his own beliefs about alarm calls, and then use this knowledge to make similar inferences about the beliefs of the food-finding species? Do they understand how the deception is pulled off, what makes it work? Does the gullible listener track the hit rates of particular antshrikes and tanagers, learning that some are more reliable than others? As the philosopher Gilbert Ryle pointed out in the late 1940s, the distinction between *knowing how* and *knowing that* is critical to any analysis of supposedly intentional actions. Animals may be very good at *knowing how* to deceive, following rules and such, but may not *know that* they are deceiving their peers. The antshrike and tanager observation shows that some animals have a mental tool that allows them to deceive. To provide evidence that animals have a mental tool that allows them to know that they are deceiving, we need other observations and experiments. Fortunately, the animal kingdom is filled with honest Joes and poker-faced cheaters.

THE HONESTY POLICY

What do the following traits have in common? Large antlers, bright-colored plumage, long singing bouts, symmetrical breasts, high leaps, and loud crying. They are all clear signs of a healthy individual. Why? Because they are relatively costly to produce and maintain. When such traits deteriorate it is typically because the individual's health has deteriorated. When they are faked, they merely function to emphasize an individual's other weaknesses. Thus, parasite-infested birds tend to have dull plumage, birds living on poor-quality territories sing less than those rich in food resources, and gazelles in poor condition rarely perform their Baryshnikov leaps when chased by cheetah. Although the first two cases have to do with characters evaluated within a species and the last example has to do with a character evaluated between species (predator-prey), the logic of costs works in approximately the same way. If a parasite-ridden bird sings its heart out, this merely draws attention to its dull plumage, a signpost for poor health. If an unhealthy gazelle is leaping about, its arabesques won't jive with its emaciated body, a clear sign to the cheetah that it is weak and vulnerable.

The biology of these systems tells us that nature has evolved an honesty policy, one based on the relative costs of building and supporting a trait. As a reasonable rule of thumb, an animal can't carry big antlers, brightly colored plumes, or a loud voice if it is infested with parasites, nutritionally depleted, or stressed from living the life of a tormented subordinate. Moreover, those individuals capable of carrying such conspicuous characteristics carry an extra cost because they are at greater risk to predation. They carry what the evolutionary biologist Amotz Zahavi called a "handicap." Handicaps are like badges, indicating an individual's capacity to skirt the heavy hand of selection, a capacity that is driven by good genes.[2]

Handicaps provide one way for an organism to evolve an honesty policy. They are signals that require no reflection, no thought, and no understanding. When an individual shows off a handicap, it is showing off a capacity to pay the costs relative to its condition. In turn, selection will favor observers that interpret them correctly as accurate indicators of condition, vigor, and ability. Exaggerated traits puzzle evolutionary biologists, however. Since they appear deleterious to survival, one would expect selection to weed them out. One answer to this puzzle is that exaggeration represents a

The handicap principle in action. Here, a hen selectively chooses the macho peacock who has survived in the face of carrying a highly elaborated set of anatomical gear, including an extremely long tail.

response to the extravagant tastes of females. Females are finicky about their mates. When they finally pick a mate, they tend, over evolutionary time, to choose males with increasingly large tails, loud voices, or brightly colored feathers.

The barn swallow is a small bird with a strikingly long, pitch-forked tail. The longer the tail, the more costly it is in terms of the aerodynamics of flight. The puzzle, then, was to figure out why natural selection wouldn't favor short tails over long tails. In the 1980s the evolutionary biologist Anders Møller set out to test the idea that long tails are favored by sexual selection. Specifically, males with long tails are better-quality males, and thus, females should mate preferentially with them. The first experiment used a procedure that experts in the field call "cut-and-paste technology." Some territorial males had their tails shortened to stubs, whereas others were lengthened beyond the norm by extra pieces pasted on with Super-glue; to control for handling the birds, another group was captured, their tails snipped, and then glued back on again. At the end of the experiment, long-tailed males clearly won, having more eggs in their nests and more young surviving as well. Long-tailed males were also better flyers, hitting fewer objects in a barn than short-tailed males. By demonstrating that they were capable of supporting the extra burden of a long tail—a handicap—long-tailed males proved their genetic worth. Although the females clearly noticed the differences in tail length, they had their eye on a second characteristic as well: symmetry.

Human eyes, orangutan arms, beetle pincers, insect antennae, elephant tusks, butterfly wings, and the pitch-forked tail of a barn swallow. A variety of animals have traits that appear on both sides of the body, often symmetrical. Sometimes, however, the developmental program that is responsible for building these characters fails, leading to random fluctuations in the position or size of a trait, one side or the other. We call these traits *fluctuating* asymmetries to distinguish them from the kinds of *directional* asymmetries that often emerge from studies of brain organization and function; for example, in most humans, the left hemisphere of the brain plays a more dominant role in language processing, while the right hemisphere is more dominant in spatial tasks. Due to the association between the expression of a trait and the underlying genetics, fluctuating asymmetries provide us with a second clause in the honesty policy. Symmetry is expressed when the developmental program works properly. In contrast, asymmetry emerges when things break down. Given a choice, females should prefer symmetrical over asymmetrical males.

To investigate the role of symmetry in female mate choice, Møller went back to his barn swallows. When males were allowed to fly inside a barn, the asymmetrical individuals bumped into the posts more often than the symmetrical individuals, and after a round of mating, ended up with females producing fewer eggs. To determine why a tail is symmetric or asymmetric, Møller traveled to the nuclear disaster site of Chernobyl. His sample included barn swallows from a local museum, specimens collected before and after the disaster. Of several dozen characteristics measured, the nuclear incident affected only tail asymmetry, and only in males. On a population level, there was more asymmetry following Chernobyl than before, and of those individuals surviving, asymmetrical males were paired with females that laid fewer eggs than females paired with symmetrical males. The exposure to nuclear waste clearly damaged the genes responsible for building a symmetrical tail. Surprisingly, it appeared to have little effect on other characteristics. The importance of symmetry in barn swallow mate choice is now joined by many more examples, including female choice for symmetrical leg bands in finches, pincer size in a beetle, spot number in guppies, and the body odors emerging from men with more symmetrical anatomy.[3]

If honesty is anchored in the animal's anatomy and physiology, then is lying possible? If it is, can animals lie by falsifying or suppressing information? If such deviousness is possible, are they aware of their actions or blind to the underlying causes?

BORN TO BLUFF

Careful observers will find mantis shrimp hidden in the nooks and crannies of the ocean reefs, peering out with their telescopic eyes. On the whole, the mantis shrimp is quite an ordinary little sea creature. However, it has one Popeye-esque claw that is far bigger than the other and has the power to break the glass on an aquarium tank. This weapon is useful year-round, but especially during the molting phase, a period of life in which the poor shrimp lies naked until a new, hard shell grows back.

The shrimps often compete over the cavities in their ocean reefs because they are limited in number and because individuals must constantly seek new locations in order to accommodate their continuously growing bodies. When an intruder approaches, the resident moves toward the entrance, making his or her presence known. If the intruder is persistent, the resident becomes more aggressive, moving out further and further, ultimately striking out with his large claw. In the 1980s the biologist Roy Caldwell uncovered an interesting twist to this story. When an intruder approaches a resident around molting time, he comes out swinging immediately, without any of the usual courtesies. The resident appears to be in a fighting mood; consequently, intruders back off. This charade continues into the molting period. Now, however, the resident is naked and vulnerable to attack. One good strike from an opponent would flatten the resident. Fortunately, the resident's apparent toughness during this period causes intruders to back off rather than pursue their attempt at a takeover.

The mantis shrimp provides us with a beautiful case of bluffing. By striking with his claw, the molting resident conveys a precise message, one designed to threaten intruders and cause them to back off. The strike represents a bluff, however, because the resident can't defend his position. The resident's bluff goes unchallenged, however, because the costs of intrusion far outweigh the benefits of cavity ownership. As with the mixed-species flocks from Peru, the economics of the situation allows residents to deceive intruders.[4]

One might expect the art of deception to reach its highest form in social species. In this light, it is perhaps surprising that mantis shrimp, asocial by nature, are such skilled liars. On the other hand, no matter how silly and vacant they may appear, chickens are highly social, and recent studies of their vocal behavior suggest that they can be extremely cagey. Individuals live in large flocks, establish pecking orders, and produce somewhere between twenty-five and thirty different calls, including of course their famous morning crow. When roosters find food, they often produce a characteristic call that recruits the hens to their side. Roosters call more often when they find really good food and are in the company of hens as opposed to other competing roosters. Sometimes they call in the presence of non-food objects such as peanut shells, which also causes the hens to approach. Like the Peruvian forest birds who produce false alarm calls, roosters make the same sound in the context of food as they do in the context of nonfood. And if the call truly refers to food, as the ethologists Peter Marler and Christopher Evans suggest (see chapter 8), then the rooster is luring the female in with a deceptive gag. By bringing her closer, he potentially increases his chances of mating.

If roosters are trying to deceive the hen, then obviously they should be sensitive to where the hen is looking. If the hen is right next to the rooster, it would be foolish to call to a peanut shell. Rather, a savvy rooster would wait for the hen to be a certain distance away and then call. In this situation, the hen can't see what the rooster has, and by the time she approaches, the object may have disappeared. By this time, the trick is in the bag. Observations show that roosters are most likely to call to nonfood objects when hens are far away, and apparently never do so in the company of male competitors.

When chickens find food, their calling behavior is sensitive to the local audience. Roosters are more likely to withhold information about food in the presence of other roosters, and call in the presence of hens, either familiar or unfamiliar. In this sense, food calls recruit sexually receptive hens, manipulating their behavior. What is unclear is whether hens that have been deceived, approaching with the hopes of finding food and finding an inedible shell instead, experience a change in their feelings toward the rooster. In fact, though we know a great deal about the social *causes* of chicken food calls, we know little about the social *consequences* of calling.[5] For deception to play a significant role in animal societies, individuals must recognize both the causes and consequences of their deceptive actions.

POKER FACES

Consider the following scenario. You are living with a close friend. It's late at night and you walk in, hungry. You head to the refrigerator and open the door, and your eyes land on the chocolate cake. Unfortunately, there is only one piece left, so you are faced with a decision. The most honest thing to do would be to call your roommate and ask whether you can polish off the last piece. Here, of course, you risk the possibility that your roommate will deny your request, saying that he was saving it for his snack. The dishonest move would involve taking the piece of cake to the closet so that you can eat in peace. Of course, withholding information is fine if you can get away with it. But surely your roommate will either catch you with chocolate crumbs on your mouth or notice that the cake is gone when he opens the refrigerator in the morning. Although you haven't actively falsified information, you have suppressed information, lying by omission.

Rhesus monkeys on the island of Cayo Santiago in Puerto Rico often call when they find food. Some of their calls are restricted to high-quality, rare foods like coconut. Different calls are restricted to lower-quality foods such as the provisioned monkey chow. The variation in rhesus food calling raises several questions. Why do individuals call on some occasions but remain silent on others? What are the consequences of announcing their discoveries as opposed to withholding such information? Are silent discoverers intentionally suppressing the information, aware that if others don't hear or see them then they won't know what has happened? If discoverers are expected to call, are they ever punished if they are caught, silent, with the goods? To address these questions, I ran a series of experiments with the ethologist Peter Marler. We presented lone individuals from a social group with a small windfall of either monkey chow or coconut. With video camera in hand, we watched the events unfold.

When discoverers saw the food, their first response was to look around in all directions, presumably searching for both potential enemies and allies. Half of the subjects tested then called, whereas the others remained silent. Because Cayo Santiago is densely populated, other rhesus monkeys soon appeared on the scene. Some discoverers were severely attacked, whereas others were either displaced without much fuss or allowed to feed in peace. One might think that low-ranking individuals would be more vulnerable to attack from high-ranking individuals who enjoy the luxury of taking food away from subordinates. Not at all. Dominant discoverers were as likely to be attacked as subordinates. Silent discoverers were, however, attacked more than vocal discoverers and obtained less food as well. They were caught with coconut or chow on their face, and paid the price.

Given that silent discoverers are attacked most, why don't they call, reaping the benefits of a few pieces of food and eliminating the costs of attack? Perhaps they are following a rule, one guided by playing the odds, one that runs something like this: "Call if you detect more dominant individuals, but keep quiet if you don't"? If this rule is correct and sufficient to explain the rhesus monkey's behavior, it suggests a kind of deception that is more like that of the mantis shrimp and chicken. Specifically, the decision to call or remain silent is simply a matter of cost-benefit analysis. There is no additional complexity in the behavior that justifies an explanation based on the discoverer's mental states, his capacity to assess his own beliefs as well as the beliefs of others nearby. The economics of the situation accounts for the patterns observed. Sometimes silence pays off and sometimes it doesn't. On a few occasions, silent discoverers escape without detection. In these situations, they eat more food than all other discoverers. On other occasions, discoverers are silent, are detected, and suffer the consequences.

A second experiment with rhesus monkeys complicates the interpretation I have just given. Together with a team of Harvard undergraduates, I ran the same kind of experiment with peripheral males, individuals that have yet to join a social group and are floating in limbo. In contrast with members of a social group, the peripherals never called when they found food, and were never attacked when caught. They either were allowed to feed in peace or were supplanted from the remaining pile of food without being chased or physically attacked. We cannot explain the peripheral males' behavior by the earlier rule. Peripheral males apparently never call when they find food. A different decision rule does, however, account for the general pattern: "If you're a member of a social group, call if you detect more dominant individuals, otherwise remain silent. If you're a peripheral male, remain silent."[6]

The rhesus monkey's behavior has all the signs of mental complexity, including strategic maneuvering, voluntary control over calling, and attribution of knowledge to others who can or cannot see and hear what is going on. We might even be tempted to say that when information is suppressed, discoverers are breaking a societal convention; when conventions are broken, punishment is the only recourse. As our discussion of simpleminded rules suggests, however, there are alternative explanations, ones that don't require individuals to infer what others believe, desire, or intend. For example, because each discoverer was tested only once, we don't know whether the variation in response is due to flexibility within individuals or to population variation that is based on differences between individuals, some adopting a strategy of silence, others a strategy of vocal bravado.

Now imagine that your roommate enters the kitchen just as you are about to reach for the chocolate cake. You obviously can't eat the cake right away, but your desires have gotten the better of you. One way to deceive your roommate is to lie about the cake, telling him that it went off, that its bad odor forced you to throw it away. If he believed you, he might actually walk away, leaving you alone, cake in hand. This type of foraging situation certainly must arise in nature, as one animal locates a piece of food before another, and then attempts to conceal his stash from a competitor following behind. But how can we shed light on the kinds of decisions that the discoverer might entertain before making his final move?

In our discussion of spatial navigation (chapter 4), we learned that chimpanzees use an optimal travel path to locate food sites that they have seen baited by an experimenter. Starting with work by the psychologist Emil Menzel in 1974, several experimenters have taken advantage of the chimpanzee's capacity to remember the location of hidden food and have conducted a competitive foraging test. One chimpanzee, a subordinate leader, watches as an experimenter baits one out of several possible sites with a banana. At the same time, a second more dominant chimpanzee (or group of chimpanzees), the follower, observes the subordinate leader, but cannot see what is going on outside. When the chimpanzees are allowed outside, the subordinate leader beelines to the correct site and eats the banana. The dominant follower can only watch the subordinate leader taking the hidden food since he lacks the requisite information to find the baited area first. Within a dozen or so trials, however, the dominant follower catches on and begins to shadow the subordinate leader as soon as she leaves the holding area. As a result, the subordinate leader loses out to the dominant follower, who takes the banana as soon as she indicates the direction of the baited site. Now, a stupid or extremely altruistic animal might continue to lead the dominant follower to the correct spot. A smart greedy animal, however, would either stop searching altogether or work out a way to dupe the dominant follower. The chimpanzee's behavior suggests that they are anything but stupid or altruistic. After losing out to the dominant follower on a few runs, the subordinate leader starts a new pattern of search. Rather than make a beeline to the correct feeding spot, the subordinate leader starts moving on a much more circuitous route. Once the dominant follower looks away, the subordinate leader takes off for the correct site. After a while, the dominant follower catches on to this move as well, and starts tracking more closely. To increase the difficulty of the task, an experimenter places a tasty banana in one location and a dull piece of lettuce in the other. Now the subordinate leader beelines to the lettuce site, and the dominant follower takes it over. The subordinate leader then moves quickly to the banana site and eats in peace. How might we explain these moves?

The dominant follower's changes are easy to explain. During the first few trials, he sees the subordinate leader run out, move to a bait site, and obtain food. This repeated pattern becomes a predictor in the same way that a light in a Skinner box predicts food for the pecking pigeon. The dominant follower thus learns that the subordinate leader is the lunch ticket. Now the dominant follower begins to shadow the subordinate leader, tracking her moves. This strategy succeeds for a while, as the dominant follower is able to take the banana away from the subordinate leader.

How are we to interpret the subordinate leader's behavior during the early trials? Given her rapid run to the baited site, can we conclude that the subordinate leader knows that the dominant follower is ignorant of the baited site because of an obstructed view? In planning her move, does the subordinate leader take the dominant follower's visual perspective into account? Unfortunately, Menzel did not run the experiments to address this question. If subordinate leaders use such information, they should be sensitive to where the dominant follower is looking, appreciating that when his head is turned or a bag is placed over his head, he can't possibly know what the observer knows.

After the first few trials, the subordinate leader and dominant follower initiate a sequence of less predictable responses. The dominant follower's change in behavior is less interesting because he simply learns a new tracking response. At first, he follows the subordinate leader to the baited site and then learns that she will meander around for a while before heading off. The subordinate leader, however, must attend to the dominant follower's actions, noting when he is distracted, watching something else. She must act like a wide receiver running fly patterns in a football game, weaving back and forth to shake off a defender. And when she is confronted with the choice between lettuce and banana, she must inhibit her immediate desires for the banana, luring the dominant follower to the lettuce first. This looks like evidence that the subordinate leader is referring to her own beliefs and desires as well as to those that the dominant follower must have in this situation. The subordinate leader must be aware that the dominant follower is aware of the subordinate leader's knowledge of the food's location. She must also recognize that the dominant follower prefers bananas to lettuce. Given this knowledge, and her own preference for banana over lettuce, she knows that she must lead the follower to the lettuce first. Unfortunately, because we have only one observation at hand, there are several alternative interpretations that we must also entertain. For example, perhaps the subordinate leader moved to the lettuce first because it was closer than the banana. Perhaps the subordinate leader noticed a slight intention movement by the dominant follower toward the banana, and thus headed to the lettuce first so that she would obtain at least some food. Perhaps the subordinate leader actually wanted lettuce first and then, when out-competed by the dominant follower, opted for the banana. Although these may seem unlikely possibilities, they must be eliminated before we accept any of the alternative explanations.[7]

GULLIBLE TRAVELS

Have you ever been lost in a big city, mapless, and in a rush to make an appointment? If so, then you know how important it is to find someone who can give accurate directions and is trustworthy. When we find someone, we evaluate their information on the basis of several characteristics, including what the person looks like, how confident they sound, and the extent to which their information appears truthful. How do animals evaluate the truthfulness of spatial information?

In our exploration of spatial navigation in animals (chapter 4), I argued that honeybees, like many other animals, have an exquisite sense of where they are in space, and where they are relative to a home base. To fly home, they use landmarks and their dead reckoning sense. They may also access, as some have argued, an internally represented cognitive map. You will recall that the biologist Jim Gould conducted an experiment to determine whether honeybees would fly out to feed from a food source located on a boat in the middle of a lake. These were the precise instructions from the hive mates who had been trained to fly out to this spot to feed on such food. After the dance, however, Gould observed that the honeybees stayed put, as if they had rejected the directions to fly and feed. Why? Because, as Gould argued, food has never been found out in the middle of this lake, or presumably any lake, and thus the information in the signal was unreliable, inaccurate. The hive members refused to move, treating the signal skeptically. The control experiment makes this interpretation quite reasonable, given that the bees flew to the boat if it was displaced the same distance away but on the edge of the lake, a presumably more likely place to find bee food.

What we don't learn from Gould's work is the nature of the bee's knowledge, the extent to which an individual's own knowledge of pollen location can override the social message. For example, if a bee knows that a field of flowers has been burned down, leaving no pollen behind, would it accept or reject a dance indicating pollen at this location a week after the burn? What about one year after the burn, when there has been time for new growth? If an experimenter brings the bees to the lake and allows them to feed from the boat, would they then follow the dancer to this location? If a bee repeatedly lies about the location of pollen, does she lose respect? Is she punished for falsely crying pollen?

We don't have answers to these questions. However, the critical aspect of Gould's work for the present discussion is that bees, and perhaps other animals, can check on the veracity of a piece of information by comparing what they are told with what they have experienced. They have a mental tool for skepticism.[8] If this interpretation is correct, then we should be able to turn reliable animals into unreliable ones.

Among East African vervet monkeys, individuals produce two acoustically distinctive vocalizations—the "wrr" and the "chutter" call—when they detect a threat from a neighboring group. In general, these two calls seem to indicate the same thing, although the chutter is often given during more threatening situations than the wrr. Having recorded such calls from known individuals, the ethologists Dorothy Cheney and Robert Seyfarth conducted a playback experiment to determine, in part, whether an animal could be made unreliable. Thus, a subject was presented with repeated playbacks of vervet Jane's wrrs until it was bored and failed to respond by looking toward the speaker. "Boring" in this case can be translated as "unreliable," for each time Jane wrred there were no other groups in sight, no competitors, no threats. Following this sequence, Jane's chutter was played back. Although the chutter vocalization sounds different, subjects failed to respond. Their response suggests that wrr and chutter calls generally mean the same thing. If Jane cries wolf about the threat of a neighboring group, then it really doesn't matter whether she uses a wrr or chutter call, she still won't be believed. If, however, one plays Sue's chutter following habituation to Jane's wrrs, then subjects respond. Although the vervets find Jane unreliable about identifying intergroup threats, Sue is reliable, at least up until the first playback. Reliability has to do with individuals.

This ingenious experiment begins to address the problems left open by Gould's bee work. We still don't know, however, whether Jane is considered unreliable for the duration of the experiment only, or whether she needs to produce a reliable call to reinstate her honesty. We can't take habituation—boredom with Jane's call—as a direct indication of unreliability. Furthermore, we don't know whether unreliability in one social situation transfers to another situation. For example, if Jane is unreliable about intergroup threat, is she also unreliable about intragroup threat, such as a challenge from a dominant male? If we bore subjects with Jane's intergroup threats, will they subsequently respond to her intragroup threats or continue to ignore her? To understand how members of a social group classify liars, we need answers to these questions.

There is an old joke that goes something like this. A guy walks into a bar and orders a beer. He drinks half of it and then realizes he needs to step outside to his car. Before leaving, he puts a note next to his glass that reads "Don't touch!" and signs it "Strongest man in the world." When he returns, the glass is empty and there is a note attached to the glass: "Thanks for the beer!" It is signed "Fastest man in the world." We see the humor because we understand the relationship between strength and dominance. We expect individuals who are weaker to leave the glass alone. We also understand why the follow-up note trumps the first one. Humor depends on people's expectations and the beliefs they set up about others. The humorous parts in a joke trip up our expectations, what we expect to follow from a sequence of events.

Dorothy Cheney, Robert Seyfarth, and Joan Silk ran an experiment with wild baboons living in the Okavango Delta of Botswana to test whether these animals would detect a violation in a natural sequence of events. This was no joke, but rather a simulation of an interaction between the barroom bully and the local wimp. Dominant female baboons often "grunt" when they approach a subordinate mother and infant. In response, the subordinate mother often "barks" back at the dominant, a gesture of submission. Thus, if you are a baboon sitting in a distant bush, you might expect to hear a dominant's grunt followed by a subordinate's bark. In contrast, you wouldn't expect to hear a subordinate's grunt followed by dominant's bark. This sequence is unexpected or inconsistent with the dominance relationship of the callers. Researchers set up playback experiments to assess whether baboons do, in fact, generate comparable expectations. As in the expectancy-violation procedures run on human infants and monkeys, Cheney, Seyfarth, and Silk expected baboons to look longer after playbacks of unnatural sequences than after consistent sequences. This is precisely how the baboons responded, looking longer when they heard a subordinate's grunt followed by a dominant's bark than when they heard a dominant's grunt's followed by a subordinate's bark. This suggests that baboons form expectations about vocal interactions based on the meaning of the call, the identity of the caller, and the dominance relationship among the interactants. They may never understand a joke, but they certainly recognize when a subordinate seems out of line.

At this point we can draw two relevant conclusions, one conceptual and one methodological. Conceptually, some animals have the capacity to detect mismatches between a signal's common function and the way it is currently being used. These animals can therefore detect cheaters, liars, and other self-interested deceivers. They have the mental tools to see through deceit. Because of this capacity, we can set up experiments that represent violations of the current state of affairs, mismatches between signal and context. This is a powerful technique for separating the gullible fools from the sharp skeptics.

FUELING THE INTUITION PUMP

In 1983, while I was a graduate student working at Cheney and Seyfarth's vervet monkey site in Kenya, I observed an unusual event, one that fueled a growing interest in the beliefs and desires of animals. The event took place among the members of a vervet monkey group living in a dry acacia woodland. Tristan, the alpha male, pursued, groomed, and attempted to copulate with Borgia, the alpha female. After several tries, including an attempt to hoist her backside into position, Tristan fumed and slapped Borgia on the head. Borgia screamed, causing all of her female relatives to come running. Within seconds, the Borgia family was chasing Tristan throughout the territory, weaving in and out of acacia trees. All of a sudden Tristan stopped, gave a loud leopard alarm call, and then sat and watched as the Borgia family fled up into the trees. Tristan remained on the ground. I searched and searched but never found a leopard or any other vervet monkey predator. Tristan's alarm call appeared, on the surface at least, to be an elegant example of deception, of actively falsifying information about the presence of a predator. But unlike the false alarm calls delivered by the Peruvian birds, this was a rare event. During a period of two years, which involved several thousand hours of observation, Tristan never performed this little trick again, nor did any other vervet monkey I observed.

Did Tristan intend to trick the Borgia family? If so, then why didn't he flee up into the trees as well? After all, if Tristan intended to deceive them, he should carry out his game of pretense to its logical conclusion, proving to the

Borgia family that he is also frightened by the sight of an approaching leopard. Perhaps Tristan thought he saw a leopard. His alarm call is thus truthful with respect to what he believed at the time, even though it was a false alarm because there wasn't, in fact, a predator at all. Like humans, animals are subject to simple error. Consequently, one wouldn't want to label someone a liar if they were simply in error. We need to know whether the Borgia family considered Tristan's alarm call an honest mistake, something that any vigilant animal is expected to do from time to time, or whether Tristan's reputation was destroyed by what was perceived as an outright lie. If they expect errors from time to time, then they should be as responsive to his alarm calls the next time around. If they think he is a liar—the boy who cries wolf—then they might ignore him on the next round of alarm calls.

In the early 1980s the psychologists Andrew Whiten and Richard Byrne began noticing that several primatologists had accumulated similar observations of what they called *tactical deception*. This form of deception is considered importantly different from other, perhaps more familiar cases such as the seductive and deadly flash of the firefly (see chapter 1) and the visual mimicry of poisonous frogs by nonpoisonous species. In these cases, deception is common, and the specific deceptive action lacks an honest counterpart. In contrast, the vervet's leopard alarm call is typically given when a leopard has been sighted. The call's normal function is to convey truthful information about the presence of a predator. Tristan's apparent use of the alarm call to deflect the heat of an angry group of females represents a new context, and as pointed out, such cases are rare indeed.

To address their hunch that primates are deceivers par excellence, and to determine whether there were differences across species in the level of deceptive expertise, Whiten and Byrne sent out questionnaires to the community of primatologists. Here's an example from the notebooks of Frans de Waal, an ethologist who has spent over twenty years studying captive chimpanzees:

> Females sometimes give away their clandestine mating sessions by emitting a special, high scream at the point of climax. As soon as the alpha male hears this he runs towards the hidden couple to interrupt them. An adolescent female, Oor, used to scream particularly loudly at the end of her matings. However, by the time she was almost adult she still screamed at the end of mating sessions with the alpha male,

but hardly ever during her "dates." During a "date" she adopted the facial expressions which go with screaming . . . and uttered a kind of noiseless scream.

The 1990 database on primate tactical deception included 253 such cases. Whiten and Byrne wish to use this database to make inferences about the distribution of deception in the primates.[9] There are, however, problems with such inferences. It appears, for example, that for some taxonomic groups such as the prosimians (e.g., lemurs, sifakas, indris) and lesser apes (e.g., gibbons, siamangs), virtually no cases of deception have been reported. In contrast, studies of wild and captive great apes (i.e., orangutans, gorillas, chimpanzees, and bonobos) provide a disproportionate number of deceptive cases. Given other studies on primates, some of which we will encounter in the next section, it seems that there is a gap between the great apes and all the other primates, the former operating with unscrupulous, Machiavellian intelligence, the latter like mindless robots.

Many of the most intriguing examples come from studies in captivity, and especially studies of individuals that have been reared in a human environment. As we discussed in chapters 5 and 6, the apes appear to differ from the other primates in terms of their capacity for imitation and mirror self-recognition, and apes reared by humans appear to have an additional enrichment of these capacities. These taxonomic patterns, however, are problematic. If we assume that evolutionary history is critical, and that species that are closer to humans on the evolutionary tree have more cognitive similarities, then we run into two counterexamples. Thus, the lesser apes (gibbons, siamangs) are more closely related to the great apes than are the macaques and baboons. Yet there are no cases of deception in gibbons and siamangs, and several in macaques and baboons. Further, gorillas show almost no evidence of deception. This absence fits with the generally poor performance of gorillas on other cognitive tests, such as the mirror mark task, and tests of imitation. Among the apes, the gorilla is a cognitive puzzle.

If evolutionary proximity fails to account for the distribution of deception, could social organization be responsible? For example, the strongest representation in the database on deception comes from species living in relatively large social groups such as the chimpanzees, baboons, and macaques, whereas the weakest representation comes from the monogamous species such as the gibbons, tamarins, and marmosets. Even here, however, we must tread cautiously, given the fact that there are cases of deception in the relatively solitary orangutan. A second potential problem is the amount of researcher-hours dedicated to studying these species. There are populations of macaques, baboons, orangutans, gorillas, and chimpanzees that have been intensively studied for thirty to fifty years, whereas studies of prosimians and New World monkeys such as tamarins, squirrel monkeys, and capuchins have rarely cracked the ten-year mark. It is thus not surprising that the longer one studies a population, the more one sees, especially when it comes to rare events.

Methodological problems aside, these single observations bring us back to an issue raised in chapter 1. Popular expositions on animals have often complained that scientists readily dismiss rare, personal observations of jaw-dropping behaviors. Here, however, we have a group of scientists promoting the power of such anecdotal observations. But there are anecdotes and there are anecdotes. For example, Jane Goodall provides a detailed account of the murderous attacks by Passion on all of Pom's infants. This is a single case study of chimpanzee brutality, but the details are extensive. In this sense, many of the cases of deception from captive apes come from detailed observations of a single animal over years and years of observation. Rejecting these single cases would be like ignoring someone who announced that they had a dog who could write computer code in Pascal, C, and Fortran. One would be foolish to ask for twenty more dogs in order to seal the case shut! One case shows that a capacity exists in the species. Understanding the distribution of the capacity within the species requires additional subjects.

In contrast with the detailed analyses of a single subject, several scientists report on rare events with little knowledge of the species and a thin description of the event. Consequently, it is more difficult to understand what happened and why. These observations are nonetheless useful because we can use them as springboards to test our intuitions about the minds of other animals.

Consider the observation that primates suppress their vocalizations when it pays to be silent. For example, Jane Goodall describes female chimpanzees suppressing their copulatory vocalizations near community borders because announcing one's presence is dangerous when a warring neighbor may be nearby. To examine the contexts in which animals either vocalize or remain silent, I teamed up with my graduate student Mike Wilson and my colleague Richard Wrangham in 1996 to conduct playback experiments with chimpanzees living in the Kibale Forest of Uganda. Chimpanzee parties travel quietly near border areas and apparently attack intruders only when they have numerical superiority. The playbacks were designed to simulate the intrusion of a single foreign male. Thus, we played back a "loud pant-hoot" vocalization to parties consisting of only females, females with one or two males, or females with three or more males. When females were the recipients, they were dead silent and either stayed put or ran off in the opposite direction from the speaker. When one or two males were present, the party stayed put half the time, but when they moved, they did so silently. In striking contrast, when parties of three or more males heard the intruder, they responded with a chorus of pant-hoots, barks, and screams, and then headed off toward the speaker, hair bristling, apparently ready to attack. These findings suggest that in chimpanzees, as in lions (see chapter 2), power comes from numerical superiority. Chimpanzees have the capacity to suppress or express their vocalizations depending on their relative vulnerability.

I draw a simple conclusion from all these observations: If you want to understand what an animal is thinking, particularly whether it is capable of deception in a rich intentional sense, then use these rare anecdotes to fuel your intuitions. These intuitions are essential for designing more careful observations and experiments.

MASKED THOUGHTS

For a brief moment, transform yourself into a female bird. The species is irrelevant, so imagine your favorite kind. You have just laid a few eggs and are in the process of keeping them warm, gathering up your reserves so that you can feed them when they hatch. While you are sitting on the nest, you spot a predatory cat moving your way. What are your concerns? Both you and your eggs are at risk. If you fly off quietly, you will be saved but the cat will discover and eat the eggs. If you attack the cat, you will put yourself at risk, but may cause the cat to flee, thereby saving the eggs. Given time, and the constraints imposed by your body plan, you can't hoist the nest to safer ground, or for that matter, transport individual eggs to a protected area. For most species, this is a dilemma indeed, and it has led to the evolution of cryptic techniques for concealing the nest. A group of birds called plovers, however, have evolved another technique, one that appears highly intentional, goal-directed, and deceptive. It is a behavioral move that has been described by ethologists as an *injury-feigning display*.

Plovers tend to nest on the ground in relatively open areas. The nest itself is cryptic, blending in with the soil and rocks. Although this form of camouflage makes it difficult for predators to find the nest, seeing a female plover would provide an important clue. When the female spots a potential predator moving nearby, she flies off in a frenzy, attempts to attract the attention of the predator, and then dive-bombs to the ground and whirls around as if she is incapable of flying. Her wing looks incapacitated, a sitting plover waiting to be eaten. The display has all the markings of intentional design: dramatic action to capture the predator's attention and convince it that there's an easy target within reach. The display also looks as if it was designed to deceive. The plover is not injured, but actively falsifies information. What, however, is going on in the plover's head prior to and during the injury-feigning pose? Perhaps she is following a rule burned into her brain, one that tells her to fly off the nest when danger is near and then run through the requisite steps of her dramatic performance. Alternatively, perhaps she sees the predator and understands that predators are likely to be attracted to injured birds and will believe that she is injured if the display is performed just so. With this knowledge, she also knows how to act injured, even when she is perfectly fine. In this sense, one succeeds at deception by understanding how one manipulates another individual's beliefs, by making them believe something that you know to be false. Deception in this sense requires the individual to be an intentional agent capable of understanding what an action is about, and how engaging in an action can alter the mental states of others, providing them with either a truthful or inaccurate picture of reality. A difficult sequence of mental steps, perhaps, but a possibility.

In the early 1980s the psychologist Carolyn Ristau set out to explore what plovers are thinking about when they use the injury-feigning display, using the natural observations described above as intuition guides for designing careful experiments. She identified a nest and parent and then used human actors as potential predators. Ristau discovered that the plovers appeared highly sensitive to where the actor was looking, as well as the path she traveled. Thus, if the actor walked near the nest and looked at it, the plovers were more likely to use the injury-feigning display than if the actor looked away. Similarly, when an actor approached and threatened the nest by stopping and staring, the plover was more likely to display than if a different actor walked by without noticing the nest. In this situation, Ristau argues, plovers discriminate safe from dangerous intruders. Does the plovers' sensitivity to the predator's behavior indicate that they are also sensitive to the predator's mental states, that they attribute particular beliefs to the predator and then respond accordingly? Do they engage in behaviors that are goal-directed, guided by an awareness of their own beliefs as well as the beliefs and desires of the other players in the game?

Ristau's work shows that the plover's display reflects more than the output of a simple rule. Implementation of the display is flexible, tuned to certain critical features of the intruder or potential predator. Thus, the plovers appear to be sensitive to where someone is looking and what the intruder has done in the past. There are even observations of plovers trying harder when a predator fails to notice them. Thus, they may swoop again, initiating the display with increased vigor. Though fascinating, these observations and experiments leave several questions unanswered. For example, though plovers are sensitive to the difference between threatening and nonthreatening objects, are they sensitive to the distinction between living and nonliving objects and if so, what is the basis for this distinction? Thus, if a beachball rolled on a path toward the nest, would they treat this object in the same way as an approach from a dangerous human actor? Would they understand that beachballs lack the capacity to move on their own and thus lack one of the critical ingredients of all intentional agents: the capacity for self-propelled motion? Do plovers recognize the difference between fruit and leaf eaters on the one hand and meat eaters on the other, understanding that the fruit and leaf eaters pose no threat to the nest? We must address these problems in order to understand how plovers represent potential threats to their fitness.[10]

If plovers are intentional creatures—guided toward goals by a set of beliefs and desires, and recognizing that others are similarly motivated—then they should be able to apply this capacity to social situations other than those that arise during encounters with predators. Our capacity to function as intentional agents allows us to lie and assess why someone else has lied, why someone desires strawberries but hates brussels sprouts, and why someone should be punished for hurting another. To date, there is no evidence that plovers use something like the injury-feigning display to fool other plovers. In fact, there is no evidence that plovers deceive in any other context but those threatening situations near the nest. This suggests that plovers are endowed with a highly specialized mental tool, one that is narrowly focused on a single deceptive context. In contrast, our capacity for intentional actions is more broadly focused, enabling us to tackle a limitless range of situations. The obvious question then is, Are other animals equipped with narrower or broader capacities for intentional action?

Normal human adults attribute mental states to others all the time. This capacity is part of our mental endowment, an ability that develops in the child without formal instruction, and allows us to infer what people desire, believe, want and intend—their mental states. We make inferences about others' mental states all the time, maneuvering by recourse to what David Premack called a "theory of mind," a set of inferences for navigating in the social world.[11] Thus, if I take the cookies out of the cookie jar and place them in the bread box, I expect my roommate to search for the cookies in the jar rather than in the bread box if he hasn't been watching my moves in the kitchen. As such, I infer that my roommate holds a false belief about the whereabouts of the cookies, and my inference is anchored by my understanding that *to see* is *to know*. Similarly, I expect that if I tell my roommate that I tossed out the cookies because they were moldy, he will believe my lie as long as my previous track record with him is clean, or if he currently has no reason to doubt the truthfulness of my claim. Lying, at this level, requires an understanding of one's own beliefs, the beliefs of the person to whom one is lying, and the tricks one has to alter beliefs.

Human infants are not born with a theory of mind. In fact, it takes a good four years for the capacity to develop. Along this developmental path, infants pick up a number of useful abilities. Well before their second birthday, young children will look to the target of a pointing finger (as opposed to the finger) and will follow an individual's gaze. Soon thereafter, they understand that what someone can see influences what they know. At about the same time, they recognize their image in the mirror, begin to experience self-reflective emotions such as embarrassment, and delight in the act of pretend play. By the end of their first year, infants understand that a stationary object can move if it is either self-propelled or moved by a second object. They also appear to understand that self-propelled objects often have goals. A recent experiment by the psychologist György Gergely and his colleagues provides an elegant illustration.

One-year-old infants watched a single video sequence, starting with a large ball stationed to the left of a small ball, separated by a barrier. The large ball pulsates and then the small ball pulsates. The small ball rolls up to the barrier and back, then rolls up and over the barrier, landing and moving to meet the large ball. The large ball pulsates and then the small ball pulsates. Once the infants were bored by repeated presentations of this sequence, they were presented with one of two test animations, both involving the same two balls, but with the barrier removed. In the first test, following the pulsation of the two balls, the small ball rolls straight to the large ball. In the second sequence, again following the pulsation of each ball, the small ball rolls forward, arcs up and then down, landing and rolling to the large ball. Given the barrier's absence, the motion of the small ball in the second sequence is bizarre. There is no *reason* for the small ball to loop up and over when there is no physical obstruction. What, then, do infants consider bizarre, more interesting, more novel? Infants consistently looked longer in response to the second sequence, a familiar path, but an odd one given the lack of a barrier.

Gergely and colleagues argue that infants look longer when an action is irrational than when it is rational. In the second test the small ball is acting irrationally. We expect the small ball to move straight ahead in the absence of a barrier, which is precisely what it does in the first test. But do infants think about this sequence of events in terms of the ball's rationality? David Premack and Ann Premack have argued that infants are designed to attribute goals to self-propelled objects, using details of the environment to infer appropriate motion. Thus, infants expect the small ball to move toward the large ball, given that this is what the balls have been doing in each of the previous sequences. Being a solid object, the barrier represents a physical obstruction to motion; thus, a particular trajectory is expected. When the barrier is removed, a different trajectory is expected. This expectation is based on physics, not psychology. Although the distinction between rational and irrational behavior is a fundamental one (see chapter 9), such labels attribute too much sophistication to the infant, especially given that we are relying solely on their eyes to inform us about their thoughts. I agree with the Premacks, then, and believe that one-year-olds use subtle variations in object motion to make inferences about goals. Human infants are not rational creatures.[12]

Around the age of three years, children begin to express and understand their own desires and preferences, as well as those held by other individuals. They also begin to show some appreciation of visual perspective, of what they can see as opposed to what someone else can see. But the three-year-old's ability to attribute beliefs to others and to understand that others can hold false beliefs is still at a relatively primitive stage. How do we know this? Present a child with a show involving two puppets, Sally and Anne. Sally tells Anne that she is going to place the ball that they have been playing with inside a box and then leave for a few minutes to go next door. Sally leaves. While she is out of the room, Anne removes the ball and places it under the bed. When Sally enters the room, we ask the child where Sally will look for the ball. Three-year-olds typically answer, "under the bed"; four-year-olds say, "in the box"; and many autistic children answer, "under the bed." The four-year-olds understand what others believe, whereas the three-year-olds and autistics haven't a clue. The four-year-olds understand that when Sally leaves the room she no longer has access to what is going on. Thus, when Anne transfers the ball, four-year-olds infer that Sally holds a false belief, and consequently, should search in the box. Three-year-olds and autistics think that what they know, everyone knows.

Here are a few twists to the false belief story, and in particular, the apparently critical shift from three to four years. In a false belief test like the Sally-Anne puppet show, let the child (rather than the puppet) move the ball in Sally's absence. Under these conditions, the age for passing this test—for correctly attributing false beliefs—drops from four to three years. If you tutor the child on problems of true and false beliefs, telling her when she is right or wrong, she also acquires an earlier understanding of the world of beliefs, who has them, what they are about, and how they change. If you have a lot of siblings, you have a richer understanding of the world of beliefs than if you are an only child. Finally, if you film the child's eyes during a Sally-Anne test, three-year-olds will look in the box where Sally should search, but then say that she will look under the bed; the kind of knowledge revealed by the three-year-old's eyes appears to be different from the knowledge revealed by her language. These tests all show, then, that comprehension of mental states can bloom before the age of four, but apparently no earlier than three.[13]

REASONED DECEPTION?

The first serious attempt to grapple with the problem of other minds in animals was David Premack's work on captive chimpanzees in the early 1970s. To determine whether chimpanzees attribute intentions to others, Premack showed his most reliable subject, Sarah, a sequence of videotapes with actors attempting to solve some problem; because Sarah loved watching television, and did so frequently, there was no problem with attention or an understanding of what the image conveyed. The sequences included an actor jumping up and down trying to reach a banana hung from the ceiling; an actor trying to push a heavy box out of his way; an actor attempting to escape from a locked cage; an actor trying to play a record from a turntable with an unplugged AC cable. Following each of these sequences, Sarah was handed two large photographs in an envelope. One of the photographs provided the solution (either an action or an object) to the problem posed in the video. Sarah's task was to select the correct photograph. For example, the correct choice for the actor locked in the cage was a key, and the correct choice for the actor jumping for a banana was to step up on a box and use a stick. If she answered correctly, the trainer said, "Good" and if she answered incorrectly, the trainer said, "Bad." Sarah picked the correct photograph in a majority of trials, and did so on the first trial. Thus, there was no possibility that the trainer's verbal encouragement reinforced the correct response. Further, she was not simply picking objects or events that happened to be associated with a problem. For example, if shown an actor shivering in a cage rather than struggling to escape, she did not pick the key, but rather the blanket. Interestingly, when three-and-a-half-year-old children were tested in the same experiment, they often selected an object that matched one seen in the video, rather than selecting the appropriate solution. Upon seeing someone reach for a banana, children picked a yellow flower rather than a box to stand on. Young children's responses were therefore guided by physical similarities (e.g., the flower and banana had the same color), whereas at least one adult chimpanzee seemed to understand that the video presented a problem, one involving an actor's unsuccessful attempt to attain a goal. From these experiments, Premack concluded that chimpanzees have the capacity to infer intentional actions.

The philosopher Jonathan Bennett raised an interesting criticism of Premack's experiments. Instead of inferring an actor's intentions from the video sequence, Sarah was merely predicting what comes next based on her personal experience, or from watching others. The crucial question then is, Was Sarah's performance based on the *literal next* or on the *intentional next*? Although Premack never examined this hypothesis with Sarah or other chimpanzees, Verena Dasser, one of Premack's students, carried out the relevant experiments with young children. Children watched a series of video sequences, each showing an actor attempting to accomplish a goal. For example, in one sequence an actor attempts to bang a nail into the wall with his bare hands, stops to look at his wristwatch, and then continues banging with his hand. In the test sequence, the same sequence is shown, but the tape is stopped immediately before the actor looks at his watch. The task: select the most appropriate *next* event out of three photographs, including an actor holding a hammer, an actor holding a pair of scissors, or an actor looking at his watch. If Bennett is right, the children should choose the photograph of the actor looking at his watch. If Premack is right, the children should pick the photograph of the actor holding the hammer. Four- and five-year-olds consistently picked the photograph associated with the intentional next—the actor holding the hammer. Three-year-olds were less successful, often selecting the literal next—the actor looking at his watch. Although it may be premature to generalize from children to chimpanzees, Premack concludes from Dasser's elegant experiments that literal next is too weak an explanation for Sarah's performance—Sarah understands intentional next.

If chimpanzees can attribute appropriate intentions to human actors, can they reflect on their knowledge of another's intentional actions, consider their own beliefs and desires, and then use this powerful system to deceive? Premack, working in collaboration with the psychologist Guy Woodruff, designed an experiment to assess this possibility, testing whether chimpanzees would indicate the location of baited food to a kind, cooperative trainer who always shared, but withhold such information from a mean, uncooperative trainer who simply ate the food once its location was revealed. Three of the four chimps indicated the location of the food to *both* trainers, thereby failing to deceive by means of withholding information. The fourth chimp eventually figured the problem out, indicating to the cooperative trainer while withholding from the uncooperative trainer. A second phase, carried out several months later and involving a few methodological changes, elicited slightly higher success: two of the chimps not only suppressed their knowledge of the baited area, but indicated the wrong area. This apparent act of deception emerged after several hundred trials of indicating to the uncooperative trainer. In a third transfer phase, the roles were reversed, and the chimps were allowed to follow the trainer's instructions. All four chimps started by following both the cooperative and uncooperative trainer. One chimp figured out that the uncooperative trainer provided misleading information after approximately forty trials, one chimp never worked it out, and the other two required approximately 150 trials. When the same test was administered almost a full year later, all of the subjects' performances improved: they indicated the location to or followed instructions from the cooperative trainer, and withheld information from or misled the uncooperative trainer.

Clearly the chimpanzees learned something, but what? Because of the lack of success on the first trial and the large number of trials required to succeed on some of the critical transfer trials, this experiment does not provide evidence that chimpanzees attribute different sorts of beliefs to each of the actors. Rather, it shows that through differential reinforcement, they can learn to discriminate between cooperative and uncooperative trainers.

Following Premack's work, experimental analyses of intentionality in animals lay somewhat dormant, whereas research on human children exploded. Then, in the late 1980s and early 1990s, experiments on captive monkeys and apes started to emerge. Cheney and Seyfarth, pioneers in this area of research, asked whether monkeys attribute ignorance to others, that is, whether they understand when another individual lacks knowledge. An animal understanding ignorance knows when to lie, when to give help to another, or when to seek it. Experiments showed that when a Japanese macaque mother was on the opposite side of a barrier from her offspring, she gave the same number of alarm calls to an approaching predator whether or not her infant could see the predator. This suggests that the mother does not take into account the infant's perspective and what this brings in terms of knowledge. Japanese monkeys, it appears, act on the basis of what others do, not on the basis of what they know. Cheney and Seyfarth summarized the field in their 1990 book *How Monkeys See the World* stating that monkeys lack a theory of mind. Though their social interactions are complex, monkeys do not attribute mental states to others. At the time, they were agnostic about chimpanzees, suggesting that more work was needed.

Following the publication of Cheney and Seyfarth's work, the anthropologist Daniel Povinelli and his colleagues designed a role reversal experiment for rhesus monkeys and chimpanzees. Although the task was identical for both species, the rhesus monkeys started out with a handicap. Whereas the rhesus were ordinary captive monkeys, the chimpanzees participating in these experiments had been heavily trained in prior work and included a kind of ape dream team, David Premack's chimp Sarah and Sally Boysen's chimpanzee Sheba (see chapter 3). The apparatus consisted of four tracks. Each track included a lever attached to a box. In phase one, the chimpanzee or rhesus operator sat on the opposite side of an opaque screen from the human informant; the screen concealed the four boxes, but not the informant's head. The operator watched as the informant concealed food inside one of the boxes. When the screen was removed, the informant indicated the correct tray and waited for the operator to pull a lever to move the corresponding box. In this first phase, it took the chimpanzees a relatively long time (80 to 240 trials) to consistently pick the track associated with the baited food. Therefore, they didn't immediately understand that the informant is like a signpost for finding the baited box. Because he witnessed the placement of food, he knows where it is. The next phase flipped the roles around, so that informants became operators and operators became informants. Two of the chimpanzees showed little difficulty switching roles, while the performance of the other two dropped. Povinelli and colleagues interpreted their findings as evidence that chimpanzees understand perspective, showing a kind of empathy in their ability to reverse roles.

There are alternative interpretations of Povinelli's results, some of these articulated by David Premack as well as the psychologist Celia Heyes.[14] Although performance may be high for a session, we need to scrutinize what happens on the first trial of a new condition, such as the role reversal condition. If, on the first trial, an individual picks a track other than the one pointed to, they could learn from this negative experience, and switch to the informant's choice on all subsequent trials. Thus, a mistake on the first trial could easily translate to success on all subsequent trials, but not because the chimpanzee understands the game. In fact, if they really understand the game, they should never miss a single trial. But they do. Overall, then, the chimpanzee's performance on this role reversal task is poor, except if we compare them with rhesus. Rhesus monkeys took about twice as long to figure out what to do in the first phase, and showed no evidence whatsoever of role reversal. In this sense, there may be a species difference. Keep in mind, however, that this species comparison tends to blur the fact that the chimpanzee dream team is competing against the poorly educated rhesus team.

To build on their results from the role reversal task, Povinelli and colleagues wondered whether chimpanzees could make inferences about which individuals are knowledgeable and which ignorant. Using the same apparatus as before and the same four chimpanzees, they started with a task including two human actors, a *knower* and a *guesser*. The knower baits one of the boxes with food while the guesser is out of the room; the chimpanzee can see the knower, but because of a screen, cannot see where the food is baited. Once the box has been baited, the guesser returns and the screen is removed. The guesser and knower then point to a box, but not the same one; the knower always points to the baited box. Two chimpanzees reliably followed the knower's selection after 100 or so trials, whereas the other two subjects took approximately 250 trials. Again, given the apparent simplicity of the task—pick the actor who stays in the room—the chimpanzee's performance is poor, and quite surprising. In the final test, the guesser stayed in the room but placed a bag over his head while the box was baited with food. This test examines whether the chimpanzees learned a useful trick—pick the guy who stays in the room—or learned something about perspective— the guy who can see knows where the food is. Three of the four chimpanzees successfully picked the knower over the course of thirty trials. However, because a lucky guess on the first trial is sufficient for the chimpanzees to learn to avoid the actor with a bag on his head, we don't learn a great deal about the chimpanzee's knowledge in this situation if we average the results over many trials. A reanalysis of the first few trials revealed that none of the chimpanzees successfully transferred their knowledge from the first phase to the second, selecting the knower over the bag-over-the-head guesser. The chimpanzees learned a useful trick. They don't understand the relationship between seeing and knowing.[15]

At a young age, children look in the same direction as someone else, understand that certain objects can be hidden by other objects, and appreciate that seeing leads to knowing. Chimpanzees and some monkeys will also look in the same direction as someone else. For example, chimpanzees will follow the line of sight of a human actor to a partition or to a point behind the partition. If they can't see anything interesting, the chimpanzees will look back to the actor, as if to check what all the fuss is about. When an actor hides a piece of food inside one of two tubes (closed on the chimp's side, open on the actor's side), or behind one of two barriers, and then stares at the baited site, chimpanzees tend to use the actor's gaze to locate the food. It seems, therefore, that they have some understanding of what an actor can see from his perspective, and use this information to make an appropriate decision.

Do these experiments show that chimpanzees understand the distinction between seeing and knowing? No, for two reasons. First, a chimpanzee can solve the problem by simply using the actor's head orientation as a signpost for where to search. A person's head and eyes can be oriented in the same direction, but they don't have to be. If chimpanzees use head orientation over eye orientation, then they certainly do not understand the relationship between seeing and knowing. To solve this problem, one would have to show that if head orientation goes in one direction and the direction of looking in another, chimpanzees preferentially attend to the direction of looking. At a neurophysiological level at least, there is evidence that some neurons in the temporal lobe of the macaque brain respond selectively to the direction of looking (independently of head orientation), whereas other neurons respond selectively to head orientation (independently of the direction of looking).[16] Assuming that chimpanzees are endowed with comparable neural circuitry, we can at least state that they have some capacity to distinguish between head and eye orientation. Second, an elaborate set of experiments on chimpanzees by Povinelli and Eddy shows an extraordinary blindness to the knowledge inherent in looking. The general setup involved training a chimpanzee to enter a room and beg for food from a human sitting on the opposite side of a Plexiglas partition, looking at them. Every time they begged, they received food. In the first test condition, one human sat to the left of a food bucket, facing the chimpanzee, whereas the second human sat to the right of the bucket, facing away from the chimpanzee. The chimpanzees begged as often from the person facing them as the one facing away. They also begged equally from a person with a bag over his head as from a person with a bag to the side of his head; from a person with a blindfold covering the eyes as opposed to a blindfold on the mouth; or from a person with an opaque screen in front of the face as opposed to the side of the face. Although these initial experiments were conducted with juveniles, performance remained poor when Povinelli and Eddy tested these same individuals several years later.

The general consensus in the field seems to be that primates lack beliefs, desires, and intentions—they lack a theory of mind. We must be cautious about this conclusion, however, given the relatively thin set of findings, weak methods, and incomplete sampling of species and individuals within a species. Even if we accept this caveat, however, some scientists argue that acquiring a theory of mind, either in development or over the course of evolution, requires language. Adherents to this view would not expect any primate to have a theory of mind. To test this idea, we must determine the similarities and differences between human language and other forms of animal communication; we must also assess whether animals have the mental capacity to learn a human language and if so, how this mental tool might alter their thoughts.

PART III

Minds in Society

All animals have evolved systems of communication. Differences lie in what animals talk about, and how listeners represent what others are saying.

8

Gossip on the Ark

All humans do it. Gossip, schmooze, chitchat, gab, talk, tattle, rap, banter, discuss, debate, and chew the fat. Why? To exchange information, share knowledge, criticize, manipulate, encourage, teach, lie, and self-promote. Since the evolution of modern humans, and perhaps before, thousands of cultures have communicated through sound, using the privileged code of human language to share their thoughts and emotions. Other parts of the body have, of course, been used as well, but history tells us that all cultures use the spoken word as the dominant medium of communication.

Human societies, at least the modern ones that we are most familiar with, seem to require spoken or signed language for their proper functioning. Prehistoric societies provide a more challenging problem, for we lack written records of their behavior and most of the materials that have fossilized say little about their communication, how it was enacted, and what was exchanged. Consequently, if we wish to explore how societies function in the absence of a language like ours, humans are the wrong species to examine. Instead, we must turn to animals and explore how their own communication systems work in the service of mediating social interactions. This statement assumes, of course, that animals lack a language like ours.

DOLITTLE'S DICTIONARY

No one has taken the art of translation further than Dr. John Dolittle, who could, as the story goes, talk to the animals. As we follow the doctor on his voyages, we learn that he readily translated the dialects of some animals, including the parrot Polynesia, the dog Jip, and the monkey Chee-chee. But when presented with the challenging shellfish, a departure from his expertise with vertebrate repertoires, the doctor was at a loss. To understand them, he recruited other animals who carried out the intermediate steps of translation.

"Dr. Dolittle" is good fiction. Many of us found his tales magical. They fueled our desire to communicate with other animals, to find out what it is like to be them. For some of us, our desires and dreams have been further fulfilled by films such as *Tarzan*, *Free Willy*, and *Babe* that either allow humans to understand animals or allow animals to speak English, at least to a selective group of listeners. If we step away from the world of fiction, can we imagine a living Dr. Dolittle, one with at least some powers to communicate with the animals, to translate their sounds and gestures to see into their minds and hearts? By decoding the meaning of animal vocalizations we could more readily understand their thoughts.

Or would we? I can recall reading the philosopher Ludwig Wittgenstein's argument that even if lions could speak, we wouldn't understand them. I was puzzled and wondered, why not? Why couldn't we use their vocalizations as a window into their thoughts? Similarly, I can recall reading the works of the psychologist David Premack, who argued that even if chickens had a grammar, they would have nothing interesting to say; again, I wondered why. Chickens, after all, live in social groups, develop dominance hierarchies, and have a large repertoire of vocalizations. Why wouldn't they have interesting things to say, grammar or not? Finally, Daniel Dennett has stated that if lions could talk, we *would* understand them, but they couldn't tell us much about their lives, whether they enjoy being predators, or what they think about the thousands of copulations required to make one cub. I find this puzzling as well.[1]

Who, then, is right? Can we translate the utterances of animals into something meaningful, something that we, as nonnative speakers, can understand? And if we can translate their utterances, do they have interesting things to say? Although animals may not understand what others desire and believe, some animals have abstract representations of object, number, space, and social relationships, and as such, have information that might be usefully shared with other members of the hive, burrow, nest, or troop. Can animal vocalizations provide vehicles for their thoughts and emotions, and more generally, the experiences they have had or will have in the future? If their natural vocal repertoires are deficient, can we tutor them, providing either a spoken or signed human language?

Consider a classic problem in the philosophy of mind and language, the problem of induction—inferring the referent of a word from a speaker's utterance. The philosopher Willard Van Orman Quine asks us to imagine a linguist who travels to a foreign land to learn the language of a newly discovered human population. One of the natives shouts "gavagai" as a rabbit runs by. The linguist's first intuition is, naturally, that gavagai means rabbit. This inference is reasonable because at least in English, when an object elicits comment, our words tend to refer to the object. Although this is a reasonable first stab, Quine pointed out that further consideration of the situation quickly brings up a sea of potential meanings for the word, including mammal, detached rabbit parts, fur ball, carrot-consuming machine, the mascot for Easter, the tortoise's competitor, the living representation of Bugs Bunny, the token for a small Volkswagen, something smaller than *Tyrannosaurus rex*, and a crucial ingredient for stew. Quine argues, therefore, that any theory of word meaning that depends on an analysis of behavior is in serious trouble. An inductive approach to meaning—accumulating facts and then drawing a conclusion—fails because there are too many possible interpretations or translations. If we can't anchor meaning in behavior, we will never work out what animals are talking about.[2]

My discussion of Quine's problem ignores other complications associated with translation, specifically, the distinction between reference and intended meaning on the one hand, and the defining properties of a word on the other. To clarify the first point, consider a waiter yelling in his order to the cook: "The ham sandwich wants an extra pickle." Although the structure of the sentence may imply otherwise, the waiter did not mean that the "ham sandwich"—some cured pork wedged between two slices of bread—wanted another pickle. He was referring to the customer who ordered the ham sandwich and his request for an extra pickle. To say that a word refers to something, then, is to say that there is a relationship between a sound and some set of objects, events, or concepts. Nothing in the sound itself tells you what this relationship is—there is nothing about the sound "ham" that tells you how to define it. Nor does the speaker have to experience the things that are picked out by the word in order to produce it—one can use the word "ham" while looking at ham, while thinking about tomorrow's lunch, or while recalling an Easter meal that made one sick. In some cases what the speaker intends to convey—what he or she means by a particular word— will correspond to the word's actual referent, but this will not always be the case. The challenge is to determine what the word refers to, what the individual intended to convey, and whether they differ.[3]

What, however, is a word? Is it a particular segment of sound, something that we can articulate in a single breath without refueling our lungs? Is it a segment of sound or writing that picks out a particular object or event, or even a concept? Some words certainly fit this definition, but others such as *the, a,* and *to* do not. Is it an abstract concept that gains meaning by its role in the grammatical structure of a language? Words certainly may play such a role, but they need not. Single words have meaning for the young child well before she strings them together into phrases. Some one-word utterances are sentence-like—"mama" may mean "mama, come here"—while others refer to a single object. Young children often spontaneously label things they see with isolated words and an accompanying finger pointed at the target of their attention. Moreover, each language defines and constrains words differently.[4]

Given the problems inherent in specifying a precise definition, let us agree that a word represents a pairing of a concept with a sound structure and associated syntactic property. A brief look at how human children acquire language helps our understanding of words and leads to a better understanding of whether animals produce word-like vocalizations.

Following a period of babbling, children begin producing their first words at around ten or eleven months, though they typically show some comprehension of words a few months earlier. Most words at this age are fairly simple and refer to specific objects, especially ones that are readily individuated—*mama, bottle, teddy,* and *kitty.* There are, of course, some wonderful exceptions to this pattern. When my colleague Richard Wrangham brought his family to the Kibale Forest in Uganda, his son David's first word after "mama" and "papa" was "colobus," the name for a beautiful black and white monkey frequently seen in his backyard. These early words tell us that children are not operating like Quine's linguist, generating a sea of potential meanings for the words they hear and then produce. Rather, the child's brain appears to limit the inferences drawn about word meaning and appropriate usage. Thus, when children hear the word "dog," they fail to make the inference that the word could refer to dog parts (tail, paw), an action performed by dogs (playing fetch, eating a bone), some setting associated with the dog (fire hydrants, trees), or the kind of substance that the dog is made of (fur, flesh). Rather, children infer that words associated with solid, bounded objects such as *dog* refer to or name specific objects. In contrast, children generate different inferences for nonsolid objects, substances such as water or sand. The grammatical structure of the sentence (e.g., *the* dog; *some* water) does, of course, help the child make appropriate inferences about the word's referent, but it is certainly not necessary.

The fact that the child's inferences about word meaning are limited or constrained does not imply that young children are born with an ability to produce flawless Shakespearean monologues. They tend to make mistakes, overgeneralizing from "Rover" the dog to all four-legged creatures. Children's mistakes show that their brains have been designed to generate a narrow range of inferences, ones that are most likely to lead to a well-organized lexicon, a mental dictionary of sorts. When children hear a new word uttered in the context of a familiar object, one to which they have already assigned a word, they tend to assume that the new word refers to a particular attribute of the object. For example, while watching an adult point to a dog and say "Look at its *tail*," children typically conclude that "tail" refers to some property of the dog. They do not infer that it refers to another word for dog, or that it is a label for a particular kind of dog.

Following the one-word stage, children experience a spurt in word learning accompanied by the ability to comprehend sentences and string words together into short sentences. These minimalist utterances read like the personal ads: *Me baby. Me male. Me cute. Me eat. Me want.* Errors of word order such as *Baby me* are rare, and not only in English, but in all languages. This shows that when the child combines words into meaningful expressions, there are constraints on this process. In parallel with the acquisition of first words, the timing and nature of the production of word strings are somewhat variable. A friend of mine likes to recount how he said little until the age of approximately three, causing some concern for his parents. One day, however, while he was in the car with his family, they pulled into the parking lot of a restaurant, whereupon he pronounced, "No beer here." Reflecting on his childhood, he says that it was just a matter of time before it was appropriate to launch a good one-liner.

How is animal communication similar to a child's path to linguistic competence? Do animals have sounds that are equivalent to our words? If so, are animals like the human child in producing her first attempts, creating sounds that paint a broad conceptual stroke? Or are they referentially narrow, picking out whole objects, object parts, events, and so forth? When they communicate, what motivates them? Do they intend to convey precise information, exchange ideas, and alter beliefs?

Fireflies flash, honeybees dance, ants lay perfumed trails, midshipmen hum, electric fish zing high voltages, lizards flash dewlaps, bullfrogs belch, chickens crow, kangaroo rats drum, horses whinny, wolves howl, lions roar, dolphins click, whales sing, baboons grunt, gibbons duet, human infants babble, and human adults talk. Each of these signals has been sculpted by natural selection for some particular function, to manipulate a group of willing listeners, to elicit cooperation, or to invoke competition. Even knowing the general function served by these signals, how are we to derive a more complete account of what the signal represents in the minds of signalers and listeners?

In chapter 5 I suggested that animals are equipped with a mental tool that allows them to distinguish between allies and enemies, kin and nonkin, male and female, old and young, and conspecific and heterospecific. Information about these categories may be encoded in their signals. All vocal species produce signals that carry information about individual identity. What evidence leads to such confident claims? On the one hand are acoustic analyses of signal morphology. In the same way that an anatomist might analyze the morphology of a hand, counting digit number and length, the expanse of the palm, and the curvature of the wrist at rest, acousticians break down complex vocalizations into their constituent parts, analyzing the duration of the signal, its frequency range, amplitude, and so on. After crunching through the calculations, we realize that unlike a human signing a letter, animals produce acoustic signatures automatically and unconsciously. I say that it is automatic and unconscious because when the individual vocalizes, its signature is included as part of the communicative package. There is no planning or calculation.

Complementing the encoding component of the signature are experiments showing that identity is readily decoded by the ear of the listener.

Thus, blackbirds, sparrows, canaries, warblers, and starlings all respond differently to playbacks of song from a familiar neighbor than to playbacks from an unfamiliar stranger. In some studies, individuals respond differently to their own song than to that of either a familiar neighbor or a stranger. In a study of vervet monkeys, playbacks of infant distress calls caused mothers to look to the speaker, but caused unrelated females to look at the mother, as if they were expecting the mother to respond.

There is some evidence that acoustic signatures are affected by kinship as well. In some primates, females from within a single maternal lineage have matched signatures. Among rhesus monkeys, for example, individuals produce a highly tonal "coo" call during friendly interactions and when attempting to maintain contact with each other. Females within a matrilineal line tend to sound more alike than more distantly related females. One matrilineal line in particular developed a distinctive signature, producing the coo with a nasal twang, sounding off like foghorns.

Kin signatures are accompanied by even higher-level identity badges at the group, population, and species level. Once again, all these acoustic signatures are produced automatically, without planning. Once acquired, they are produced as default values, offered to anyone who will listen. Although the environment will certainly distort the signal to some extent, the task of decoding at this level is relatively straightforward. Typically, the animal can hear the signal and work out species, population, group, sex, and individual identity from that information.[5]

The recognition problems discussed thus far are relatively simple. They would, however, stump even our most sophisticated computers. Although one might think that a simple pattern-matching program would do the trick—build some templates and then match them up to the input—this is currently impossible. Such complexities aside, we haven't yet discussed whether animal signals have any specific content. When an animal calls, does the vocalization refer to or pick out any particular emotional state, object, or event?

In birds and mammals, young produce characteristic vocalizations when they have lost contact with their mother or when their attempts to elicit care have failed.[6] What, if anything, does the crier intend to represent with its cry and what, if anything, does a listener think when it hears a cry? Early in life, the cry seems to be a reflexive response to emotional distress, a response that is as automatic as the human knee jerk. Young animals cry when a particular situation arises, without planning, volitional control, or conscious goals. Rather, there is a functional connection between the emotion experienced, the context eliciting the emotion, and the signal generated. In other words, when an individual is in pain or hungry, it produces a sound with a specific structure. This nonarbitrary connection between context and signal structure stands in striking contrast with human words. For example, when we use language to describe the sound used by an infant in pain, we say that the infant is "crying." The word "crying" is arbitrary in the sense that there is no necessary connection between the context and the sound used to describe it in words. We could just as well use the emotionally opposite word "cooing" or the French "pleurer." The power of our words comes from the arbitrary nature of the association between sound and meaning. If we agree that a particular sound carries a particular meaning, then it will. Our capacity to remember and articulate the sound limits us somewhat. Our emotional expressions, and certainly many comparably designed animal expressions, are not so malleable. Dogs, for instance, appear to lack the capacity to shift gears and use their bark to indicate that they would like the filet mignon sitting on the kitchen table, rather than barking to indicate danger.

Over the course of development, the child gains increasing control over her cries. She becomes sensitive to context, to who is watching and listening. If she falls and hurts herself when no one is around to comfort her, she may sidestep the crying routine, while if someone is watching her, she may cry at length. Although the child can now signal intentionally, voluntarily conveying a message to willing listeners, the cry does not refer to a specific event or object. A parent listening out of sight would know that something has gone wrong, but would not know what caused the problem.

During my studies of rhesus monkeys, infant monkeys would occasionally sneak up behind me and then, without provocation, begin screaming. This theatrical performance readily recruits older and stronger rhesus who come rushing in, prepared to defend the apparently threatened infant. They of course did not know what happened before their arrival. All they see is an apparently petrified infant crying while looking up at poor innocent me. The vocalization has been paired with an emotion that roughly corresponds to being afraid. In the absence of any more contextual information, our translation efforts—and those of the animals listening in—can go no further. Here, then, the relevant information is carried by context, by what is happening right then and there.

What about the listener? When parents hear someone crying, what is represented in their head? If pain and hunger cries elicit different responses from caretakers, what does that say about the decoding process? Crying would seem to refer to a particular class of emotional states, but if we think of the problem in the following way, we realize the erroneousness of this interpretation. Assume that the infant's face turns bright red when she is in pain and turns bright green when she is hungry, alterations that are due to changes in blood flow. As with heart rate and pupillary dilation, individuals have no control over blood flow to the face. However, parents readily learn the salient cues from their infant's cries in the same way that they learn about traffic lights, green for go, red for stop. The fact that attentive individuals learn to discriminate by associating stimulus and consequence does not help us understand why the stimulus was produced and what, if anything, it refers to. Yet for many animal vocalizations, all we have are observations of a signal followed by some kind of response, and sometimes the response is nothing more than a rapid glance toward the signaler. When one individual responds in a highly specific way to another's vocalization, this need not imply that the vocalization refers to a particular object or event.

A WORD IS BORN

When people work with animals for a long time, the animals often appear in their dreams, either as observers on a voyage or as sages holding deep secrets. While I was watching vervet monkeys in Kenya, I dreamed about an infant who held *The Book of Vervet Translations*, a work filled with definitions and common phrases. At the time, I was intrigued by a vocalization called the "wrr." Among adults, the call is given when one individual perceives a threat from a neighboring group. Infants, in contrast, tend to produce the call when they have been separated from their mothers and are distressed. In my dream, an infant named Unique New York (like her peers, she was named after a tongue twister) came up to me and started wrr-ing, over and over again. I tried asking her to explain the meaning of wrrs. She simply wrr-ed back. I also asked if I could look at the book, but to no avail. I patiently waited for another dream, one with greater illuminative powers. No luck.

My dream captures two important features of translation. The first is the problem of meaning, the challenge of decoding both a word's referent and the speaker's intended usage. The second is the problem of segmentation, the ability to pull out words, phrases, and sentences from a continuous stream of sound. When we listen to our native language, the words appear to pop out, effortlessly. Our brains are equipped with a mechanism that finds word boundaries and allows us to extract meaning from the acoustic stream. If you were to look at a visual representation of sound—what acousticians refer to as a "sound spectrogram"—you would be hard pressed to circle the words. Your first intuition might be to look for local decreases in the amplitude or intensity of the signal, using such pauses to demarcate where one word ends and another starts. Sometimes you would be right, but sometimes you would be wrong, placing a demarcation point within a word. The mistake arises because there are often physical breaks between the sound units, or phonemes, that make up a word. To illustrate this problem, try saying the phrase "please stop." If there's any break, it's after the "t," not after "please," where there's a continuous "zs" sound. The problem of segmentation becomes even more difficult when we listen to a foreign language, especially when the language is a distant cousin to our own native tongue.

Adults who restrict their linguistic efforts to the native language experience the segmentation problem once in their lives. During development, all human infants are confronted with speech, but no speaker of the language stops to tell them where the words are. Nor do native speakers necessarily exaggerate the words and pauses to make the task easier for the linguistic pupil, though all of us occasionally raise our eyebrows, point to an object, and say, "Look at the *baaallllll* " to an infant. While people speak, infants listen, read the speaker's intentional use of noise, and ultimately emerge with a mental lexicon. The pace of acquisition is phenomenal: infants tack on about nine new words a day from eighteen months to six years of age. Although there is no consensus on how infants work out the segmentation problem, some researchers suggest that within a given language, there are statistical regularities regarding the frequency with which the combination or sequences of sound segments appear. Thus, for example, *st* is a common consonant string in English, whereas *gd* is not. Similarly, "ng" is always syllable-final in English, whereas it can be syllable-first in Swahili. Such regularities are likely to help the infant carve the native language into functional components.[7]

The segmentation problem is equally challenging for humans listening to animal sounds and for young animals attempting to acquire their own species-typical vocalizations. Scientists who study animal communication tend to approach the problem with two unfortunate biases, our own hearing system and a human-centered view that attempts to use linguistic structure to decode other signaling systems. Our hearing system is comparable to those of most mammals, but certainly not all, and it is certainly different from that of birds, fish, reptiles, and insects. For example, the range of frequencies that we hear is no different from a chimpanzee's. It is, however, wildly different from that of microchiropteran bats that hear into the ultrasonic range necessary for processing their biosonar signals, or elephants and whales that hear into the infrasound range, making it possible for them to transmit signals to willing listeners over several kilometers.[8] Compounding this difficulty is our human-centered point of view, which often leads us to dismiss vocal systems as uninteresting if they lack a rich system of reference or a combinatorial system that enables the individual to recombine a finite set of elements into an infinite variety of expressions. Although this chapter centers on the relationship between animal communication systems and human language, I want to make clear that there are many other aspects of animal communication that have nothing to do with human language.[9]

Animal repertoires vary in size from one or two distinctive signals up to twenty-five to thirty. Within these repertoires, we find signals that are used during dominance-related competition and cooperation, aggressive fights with territorial neighbors, food discovery, predator detection, mate attraction, group movement, and play. The earliest descriptions suggested quite neat packaging, one call associated with one context. A simple vocabulary, with an equally simple set of definitions; this turned out to be an oversimplification, however.

In the late 1960s and early 1970s, just around the time when several psychologists were attempting to teach apes and dolphins some form of human language (see the section in this chapter entitled "Schoolmarms for Hire"), a group of field biologists began studying primate vocal repertoires under natural conditions. These studies revealed that primate vocal repertoires were more variable than had previously been described, in terms of both their acoustic morphology and the diversity of contexts in which calls were produced. Importantly, there were strong hints that primate vocalizations were more than "groans of pain," to borrow a phrase from the biologist Don Griffin. Some vocalizations seemed to be word-like, referring to particular objects and events in the environment such as a predator, some food, a sexually receptive female, and so forth. Struhsaker's work on vervet monkeys is perhaps the most exciting in this light, especially because of the impact it would have on future research, setting up a revolutionary perspective on animal signals. By the time Struhsaker and his successors had finished their work, the classical view of animal communication was just that, classical but no longer credible as a complete account.[10]

Vervet monkeys are Old World primates, close relatives of the baboons and macaques. Due to their size and the habitats they have invaded, vervets often fall prey to a variety of predators. In Amboseli National Park, located in the southern part of Kenya, Struhsaker reported that the vervets were vulnerable to attack from martial eagles, crowned hawk eagles, leopards, cheetah, jackals, baboons, humans, black mambas, green mambas, and pythons. The poor vervet would seem to have little chance of surviving such intense predation pressure. Natural selection, however, operates on both predators and prey, resulting in an evolutionary arms race over time. The outcome of this arms race is that a highly specialized, domain-specific mechanism for responding to predators has evolved in vervet monkeys.

Most prey species have evolved a single, general-purpose alarm call that functions to alert close kin or to warn the predator that it has been detected. For vervet monkeys, however, a general-purpose alarm call would be disastrous, given the differences in hunting style across predator types. Apparently in response to such variation, vervets have evolved different escape strategies and associated warning calls for each of the primary predator classes; to date, most of the focus has been on the large cats, eagles, and snakes, although vervets also give different-sounding calls to small carnivores, humans, and baboons.

Leopards and cheetah remain quiet and hidden, waiting for a potential target to move within range and then darting after it. When vervets see one of these cats or hear the alarm call that is typically paired with them, they respond by running up into trees, or if already in a tree, moving up even higher. Though leopards and cheetah can and do climb trees, they are unable to reach the fine terminal branches that support vervets. The sound of the cat alarm call is noticeably different from the eagle alarm call. The latter is given to warn off martial and crowned hawk eagles, which soar or swoop from the sky. Upon hearing such calls, nearby listeners look up, scan, log the eagle's coordinates, and then head for the nearest bush. In contrast to the big cats, eagles can readily pluck vervets from a tree. Consequently, a dense bush provides the greatest protection from an eagle attack. Snakes represent the third major predator class. When vervets spot a mamba or python or hear a snake alarm call, they stand bipedally and search the ground nearby. Like the big cats, snakes hunt by sitting and waiting. The vervets' response is therefore geared toward detection rather than escape. Once the snake has been detected, the best option for a vervet is to stay put, keeping an eye out for possible movement in its direction.

These observations raised the possibility that vervets had evolved a system of symbols for each predator class, word-like sounds that referred to specific types of predators. This theory was supported by a more careful analysis of the signal as well as the response. For example, the cat alarm call is a loud, multisyllabic signal, apparently designed to inform both the predator and other group members. The eagle alarm is also loud, but a single syllable that is more difficult to localize, suggesting that it has been designed for the vervets' ears rather than the eagle's. Finally, the snake alarm is a quiet, multisyllabic signal. From a design perspective, this makes sense since snakes can't hear and the only relevant audience consists of those vervets sufficiently close by that they might walk into the danger, precisely what snakes anticipate.

The vervets' responses to these alarm calls provide a second means to understand how they classify or represent each predator class in their brain. The fact that individuals scan the horizon in response to cat alarms, the sky in response to eagle alarms, and the ground nearby in response to snake alarms suggests that they must have some kind of search image in mind, some kind of predator representation that they are seeking to match. Although intriguing, such observations failed to clarify what the vervets were thinking about when they heard such calls.

This was the state of affairs in 1969. Further clarification would wait until the late 1970s, when Robert Seyfarth and Dorothy Cheney joined Peter Marler on a project to explore further the meaning of vervet alarm calls. They hoped to test an explicit prediction from Struhsaker's work. If the vervets' alarm calls were like words—sounds that refer to particular predators—then listeners should be able to respond appropriately to them even without any contextual information. In the same way that I know to run out of a building if someone yells, "Fire!" vervets should know which escape response to select when they hear the cat, eagle, or snake alarm call. We don't need to see or smell smoke, and vervets shouldn't need to see a predator or see others fleeing in a particular way.

Seyfarth, Cheney, and Marler first recorded alarm calls from known individuals in known contexts. With a library of sounds on tape, they played back calls through speakers concealed in dense vegetation and filmed the vervets' responses. Subjects responded to these playbacks as if an actual predator had been detected. Upon hearing an eagle alarm call, for example, subjects scanned up and then headed for safety in a bush. These distinctive responses suggest that hearing an alarm call evokes a representation of a kind of predator.

Seyfarth, Cheney, and Marler concluded that vervet alarm calls are referential, functioning like some of our words: "leopard," "eagle," and "snake." This was a revolutionary claim. It replaced the dominant view of animal communication, suggesting that at least some animal signals convey both emotional and referential information. Further, it suggested that humans share with other animals a crucial feature of language—arbitrary sounds that have referential content.

Scientific claims that overthrow a dominant position are always closely scrutinized. Thus, a barrage of questions, criticisms, and alternative explanations emerged. Some of the criticisms were readily dismissed. Others required additional empirical work. Still others remain as substantial problems today that may not be resolved any time soon. Let me flesh out some of the more fundamental issues.

When vervets produce alarm calls, there is no question that they are afraid. In fact, different individuals appear to experience different levels of fear due to differences in their vulnerability to cats, eagles, and snakes. Adult males, for example, call more for leopards than they do for snakes and eagles, apparently because they are more likely than adult females and juveniles to be attacked by a leopard. When a leopard is first sighted moving nearby, the intensity and rate of calling are high, but the call is nonetheless identifiable as a cat alarm call. As the leopard moves on, and thus, the level of threat declines, the call remains the same, but becomes less intense and frequent. If we take the level of threat as a measure of the vervets' level of fear, then this situation reveals that certain components of the signal convey information about emotional state, while other components reveal information about the external threat. Vervet alarm calls are therefore vehicles for both the caller's emotional state and the object encountered. These observations are mute, however, with respect to what the caller represents in his head when he utters an alarm call.

In Amboseli National Park, only 30 percent of all vervet infants survive past the first year of life. Most of the mortality is due to predation. Under such conditions, one might expect natural selection to favor rapid acquisition of a fully functional alarm call system, including, from birth, the knowledge to use alarm calls appropriately as well as to respond appropriately to them (chapter 6). In the real world, however, something different occurs, something typical of relatively long-lived animals that spend their lives building up a storehouse of information. Although vervet infants produce adult-sounding alarm calls from birth on, it takes close to a year for them to learn how to respond appropriately to these calls, and even longer for them to learn how to use them in the appropriate context. Thus, upon hearing a cat alarm call, young infants inappropriately dart out into the open, freezing like sitting ducks. Regardless of the representational properties of vervet alarm calls, infants have yet to acquire the information needed to respond appropriately to them.

The development of alarm call usage is perhaps even more telling, revealing how innate factors limit the earliest development of the system and then determine how experience adds the finishing touches. When infants under the age of one year see an object falling out of the sky, they give an eagle alarm. The diversity of objects that elicit this call is broad, however, including nonthreatening vultures, pygmy falcons, and falling leaves, as well as martial and crowned hawk eagles. Critically, infants never produce an eagle call for objects moving on the ground. Conversely, although infants produce the cat and snake calls for a large number of objects, many of which are nonthreatening such as a stick dropping down through a bush, they never produce these calls for objects moving above ground. Such observations reveal an innate association between the three alarm call types and the three general classes for objects and their spatial locations. The range of objects falling within each class is large at first, and then reduced as a function of experience. In the same way that a human child may use the word "dog" to refer to her pet poodle and all other four-legged creatures, it appears that vervets do the same for their predators. With experience, the developing vervet monkey refines the class of objects or events to which a call is applied.

The vervet alarm call system shares at least three properties in common with human words, particularly with what linguists refer to as "count nouns," words associated with objects that can be individuated. First, there is an arbitrary relationship between the referent and the structure of the sound. Cat alarm calls don't sound like leopards, nor do eagle alarm calls sound like eagles. Although the vervets are certainly afraid of these predators, their level of fear does not determine the kind of call used. Second, the vervet monkey's brain does not predetermine the class of objects or events that are associated with a call; rather, vervets acquire this association as a function of experience. For example, vervets living in Nairobi, where leopard and cheetah are absent, produce the cat alarm call for domestic dogs, suggesting that the call may refer more generally to large land predators. Third, vervets can determine the referent of the call in the absence of seeing the caller or the events responsible for eliciting the call; the acoustics are sufficient when an animal knows the code.

Given the similarities identified, are vervet alarm calls like words or aren't they? Like most straightforward questions, this one is in fact difficult to answer. It refers us to a problem that we have repeatedly encountered in the previous chapters, the problem of similarity and the extent to which a suite of shared characteristics can be considered evolutionary precursors or not. If we found a fossil mammal with a forelimb consisting of a palm and three digits, would we be satisfied in classifying this structure as the precursor to a modern human hand? What if it had one digit, and a short stubby one at that? As morphologists and systematists have discussed for centuries, identifying precursors depends on analyses of both structural and functional similarities. Functions don't fossilize, however, though one can make reasonably accurate inferences from the biomechanical properties of a structure. Thus, one way to answer the question about vervet alarm calls is to consider not only structural similarities, but functional ones as well. The vervets' alarm calls function to alert group members of the presence of a particular type of predator. In this sense, we might also say that a human infant's cry is functionally referential, designed to convey information to caretakers that it is in a certain emotional state, one requiring comfort or food. Both vervet alarm calls and human cries are unlike human words in that knowing their acoustic properties allows one to determine, precisely, what the caller experienced. In contrast, if I yell "eagle," it might be that I have just spotted an eagle soaring above, or that I anticipate seeing one, hope to see one, or would like to eat one. "Eagle" picks out a relatively unambiguous object, even though my intended meaning is unclear.

• • •

When a honeybee returns from a foraging trip, it communicates information about the location of food in a dance that indicates whether the food is near or far, to the south, north, west, or east, and whether it is of high or low quality. Such information must be in the dance so that hive members that attend and then leave the hive can find the appropriate feeding spot quickly and easily. In fact, the biologist Axel Michelsen demonstrated in the 1990s that honeybees will follow the instructions from a robotic bee doing a dance. In addition to these quite explicit instructions, the bee's dance adds on a design element not present in the vervets' alarm calls. While vervet alarm calls are produced only in the presence of a predator—excluding the few cases of apparent deception that we discussed in chapter 7—the honeybee appears to refer to an object that is displaced in time and space. Even here, however, we must temper our interpretation because we lack critical pieces of information. We don't know, for example, whether the dancer is saying, "I just encountered some great food six hundred meters away to the south" or "You should fly off six hundred meters to the south to get some great food." The first refers to the bee's prior experience, a kind of report on a recent event, whereas the second implies the bee's prior experience but explicitly states a set of instructions for what to do in the near future. Nonetheless, the bee's dance is not as tied to context as are the vervet alarm calls. The honeybees' apparent capacity for reference is restricted to only a few topics. They can chat all day about food, but not much else. In this sense, the bees' signaling system represents another example of a narrowly focused capacity.[11] In contrast, additional work on vervet monkeys, other primates, and even the domestic chicken has revealed that the capacity to produce signals that refer to objects and events in the environment is more general, popping up in several social contexts. Surprisingly, perhaps, there is no clear example of such referential signals in the great apes, though there is some suggestive evidence that the chimpanzee's food grunt may represent a candidate example.[12]

Given the variety of call types in a vocal repertoire, do animals classify vocalizations in terms of their acoustic or referential similarities? We know that animals form acoustical categories based on their natural responses to vocalizations and their ability to use operant responses to classify sounds according to call types. But what determines the categories? Sometimes words sound the same, but mean quite different things, such as lecherous and treacherous. Conversely, some words mean the same thing, but sound quite different, such as treacherous and unfaithful. Are animals sensitive to such distinctions, and do they play an important role in perceptual classification?

Why do rhesus monkeys produce five acoustically distinctive calls in the context of finding food? Do they all mean the same thing? When they hear the calls, what determines the process of classification, acoustic or referential similarity? Rhesus produce three call types, the "warble," the "harmonic arch," and the "chirp," when they find rare, high-quality food such as coconut. They produce two call types, the "coo" and the "grunt," when they find common, low-quality food items such as Purina monkey chow. To explore whether rhesus monkeys classify the warble, harmonic arch, and grunt on the basis of acoustic or referential similarity, we used a habituation procedure. Here's the logic. A mother is cooking in the kitchen while her young daughter desperately tries to attract her attention, repeating over and over again, "Mommy watch . . . Mommy watch . . . .Mommy watch . . . Mommy watch . . ." Mom responds to the first request, saying, "That's nice, dear," but then returns to cooking. On subsequent repeats, Mom simply nods and says, "Yes, that's nice." The little girl suddenly alters her tactics and says, "Jane watch . . ." Although this sounds different, it translates to the same thing, the same referent, the same person—Mom. Mom continues attending to her cooking, repeating, "Yes, that's nice." Realizing that she has failed to capture her mother's attention, she says, "Daddy watch." This of course grabs Mom's attention, not because "Daddy" and "Mommy" sound different, but because Daddy refers to something different. Mom now turns to look at her daughter, and presumably, to find out whether her husband has just walked in. Mom has, quite simply, shifted from being bored to being interested because of a meaningful change in her daughter's words.

I presented rhesus monkeys with a Mommy-Daddy type test. I repeatedly played back one call type until the subject stopped looking in the direction of the speaker; then I played back a different call type. The first sequence involved a comparison between warbles and harmonic arches, each produced by the same individual. Although these calls sound different, rhesus produce them in the same general context. If classification depends on acoustic differences, then following habituation to warbles, listeners should dishabituate or respond to harmonic arches; similarly, subjects should respond to warbles after habituating to repeated presentations of harmonic arches. In contrast, if classification is guided by the call's referent, then following habituation to one call type, subjects should remain uninterested and bored when they hear the second call type.

Rhesus monkeys responded to changes in the call's referent and not to the acoustics. Having habituated to warbles, they failed to respond to harmonic arches; similarly, they made no response to warbles after listening to a sequence of harmonic arches. This suggests that for rhesus monkeys, warbles and harmonic arches form a single category, even though they sound different.

The next sequence of playbacks compared call types that differed in terms of both their acoustics and their potential referents. In one condition, rhesus monkeys heard either warbles or harmonic arches until they were bored, and then heard a grunt in the test trial; in the second condition, rhesus heard repeated playbacks of grunts until they were bored, and then heard a warble or harmonic arch in the test trial. One problem with this design is that a response in the test trial could be due to either acoustic or referential factors. There is, however, a solution to this conceptual pickle, revealed by the rhesus data. Although subjects consistently responded on the test trial by looking toward the speaker, they looked longer in response to warbles and harmonic arches than to grunts. In other words, although each playback test contrasts the same call types, the strength of the monkeys' response in the test trial was affected by what they heard before. If you have been eating caviar for a long time, a plate of boiled potatoes is unlikely to be that interesting, whereas caviar following potatoes may well capture your interest. The rhesus monkeys seem to be telling us that a shift from grunts to warbles or harmonic arches represents a meaningful change, one that reflects a change in the call's referent. In contrast, when a caller shifts from warbles or harmonic arches to grunts, the referent has changed, but such change is not that interesting.

Like rhesus monkeys, other primate species such as vervets, baboons, and diana monkeys also classify vocalizations on the basis of their apparent referents; these studies have explored contexts other than food, including predation, intragroup friendships, and intergroup aggression.[13] Unfortunately, we still have only a vague understanding of what these vocalizations refer to. Why, for example, do rhesus monkeys have three calls for high-quality, rare foods? Although it seems unlikely, perhaps these calls are synonyms, something like "food," "grub," and "chow." Alternatively, these calls may be associated with different food types in more natural rhesus habitats, or they may reflect different levels of excitement upon finding high-quality, rare food. To clarify these issues, we need to run additional experiments, manipulating the animal's level of excitement as well as the kinds of food presented.

Our current understanding of animal communication suggests that human words and animal calls are based on quite different mental tools.[14] Animal calls generally indicate things in the here and now. Human words indicate things in the here and now, but also in the distant past and well into the future. When we observe an animal's behavior, and especially the objects and events it encounters, we can predict with a high degree of certainty which calls it will produce. When we watch a human's behavior, our ability to predict his or her words is relatively poor. When a vervet monkey encounters a leopard and is in the presence of other vervets, we know precisely what the call will sound like. When a human encounters a leopard, he might say, "Wow, how beautiful!" "A leopard!" "Chui" (Swahili for "leopard"), or "That's no kitty!" Whereas animal calls pick out concrete objects and events in the animal's environment, human words pick out much more, including abstract objects and events and whatever our imagination can generate. These differences make it difficult for scientists to trace the evolutionary origins of human words back to an animal precursor. Rather, a majority of scientists conclude that words, as we know them today, originated somewhere along the human branch of the evolutionary tree.

Philosophical analyses, together with several studies of child language acquisition, tell us that the ability to refer is different from the ability to communicate intentionally. One could well imagine building a robot capable of referring to objects and events in its environment, but failing to act like Hal in Stanley Kubrick's *2001: A Space Odyssey*, a computer with intentions, desires, and beliefs. Can we take, however, what the philosopher Daniel Dennett calls the "intentional stance" and infer from an animal's vocal behavior that it is sensitive to its own beliefs and the beliefs of others nearby.[15] Do animals have voluntary control over their signals, such that when they communicate, it is with the intent of achieving some goal?

Some would claim that an animal communication system with reference but without an underlying system of intentionality is nothing like human language. Thus, are vervet monkey alarm calls more like traffic lights than traffic guards raising their hands to control the flow of cars and trucks? Traffic lights tell others when to stop and go, but if a traffic jam occurs, they go on signaling in the same way, oblivious to the current disaster. Traffic lights can't suppress their signals or modify when they use them. Traffic guards can.

The catalyst for examining this problem is a small sample of onetime observations, some of which we considered in the last chapter. In particular, Cheney and Seyfarth observed a few cases where a lone vervet monkey, out of sight from other group members, encountered a predator but failed to call. This suggests that vervets are sensitive to their audience, intentionally producing an alarm call when the target audience is near, but suppressing or inhibiting the call when it is not. Similarly, there were several cases where vervets produced an alarm call in the absence of a predator, but in a situation where the individual's simulation of a predator attack caused an aggressor to abandon the chase. Cheney and Seyfarth later conducted an experiment with captive vervets living in the San Fernando Valley of California, showing that adult females were more likely to give alarm calls in the presence of a human predator when they were with their juvenile daughters than when they were with unrelated juvenile females. Together, these data suggest that vervets have voluntary control over their alarm calls, and their production is guided by the intent to convey information to particular listeners.

The observations on vervets, combined with some of the aforementioned observations of audience effects in chickens and rhesus monkeys, clearly show that animal vocalizations are not reflexes. They are not automatic or involuntary. But are they intentional? When a vervet monkey gives an eagle alarm call, does it want listeners to know it believes that an eagle is hunting and that they should run for cover? In combination with the work reviewed in chapter 7, current observations suggest that animal vocalizations are not guided by the beliefs and desires of their friends, enemies, or kin. Japanese macaque mothers, for example, do not modify their alarm call responses on the basis of their infants knowledge of an approaching predator. They seem blind to what the infant knows. In this sense, animal vocalizations are not intentional even though they have been designed to achieve a specific goal.

NATURE'S GRAMMARIANS

René Descartes and Noam Chomsky would agree on at least one important point: Humans are equipped with a grammatical system that enables them to string words together to form a limitless variety of meaningful expressions. Animals lack this system. As Descartes put it almost 350 years ago, "It is a very remarkable fact that there are [no humans] . . . without even excepting idiots, that they cannot arrange different words together, forming of them a statement by which they make known their thoughts; while on the other hand, there is no other animal, however perfect and fortunately circumstanced it may be, which can do the same." It appears, then, that humans are equipped with a specialized mental tool, one that enables us to produce a limitless range of meaningful expressions. Is this claim correct?

When songbirds deliver their songs, they arrange the notes in a specified order, providing information to others about their sex, motivational state, geographical location, and species. Young birds reared without hearing song produce a song in adulthood that lacks the proper species-typical signature, including the well-ordered sequence of notes that are characteristic of the species. If a disturbance occurs in the midst of a singing bout, the singer stops before completing the song, but only after producing one of the themes within the song. There are, in essence, natural breaking points that spotlight salient units within the song.

The motor actions that the songbird uses to produce song are based on rules for ordering motor sequences. For songbirds, innately specified mechanisms set up the rules for imposing order on a sequence of elemental units, but the particular circumstances in which the individual finds itself will fine-tune the details.

The rules for sequencing motor actions involved in song production appear to be similar to the rules that structure all human languages. But similar in what way? To explore this question, not only must we assess what kinds of rules animals follow in structuring their signals, but we must also explore the extent to which such rules permit a relatively open or closed system of expression. To what extent can animals expand the range of meaningful utterances they produce by stringing calls together? Are they limited to labeling or naming discrete objects and events in their immediate environment—such as leopard, coconut, attack, approach—or can they put calls together into a sequence that captures more subtle nuances of the situation, referring to the size or number of leopards or pieces of fruit, and whether they are near or far? If these enhancements are possible, do their calls fall into such abstract categories as noun, verb, and adjective?[16]

In addressing the questions above, we face two general problems. First, we must work out how animals carve up the sound space that their vocal production and perception systems permit. In terms of units, is the call equivalent to the word? Can calls be combined, and if so, in what way? Do the units combined maintain their meaning or are they blended, as when white and red paint mix to create pink? The rules that operate in human language are similar to the rules that operate in genetics, chemistry, and math, a point articulated by the theoretician William Abler and applied more directly to problems in language evolution by the psycholinguist Michael Studdert-Kennedy. In particular, the expressive power of human language comes from the fact that it recombines a finite set of elements into a variety of novel and meaningful sequences; blending systems, in contrast, fail to create such expressive power because they average out the potentially interesting meaning or variation that each element carries on its own.

The second problem is whether different sound sequences mean different things to speakers and listeners. When we string words together, the order in which the words appear is critical to the intended meaning. Well before human children are competent sentence users, they are sensitive to word order, using it as a guide to make proper inferences about what a speaker means when she produces a sentence. When a two-year-old hears "Big Bird gives Elmo a cookie," she will look at a picture depicting this action sequence as opposed to an alternative showing Elmo giving Big Bird a cookie. Although the words are the same, the way they have been recombined captures something essential about what the speaker means. The evolutionary problem, then, is to uncover the kinds of pressures that would have favored a particulate system of communication and when, over evolutionary time, such pressures crafted the kind of system that we see in modern humans.[17]

From a purely structural perspective, one focused on the morphology of sounds and sound sequences, the courtship songs of birds, whales, and gibbons share more in common with human sentences than the shorter grunts, screams, barks, clicks, chirps, and trills that accompany their other social interactions. When humpback whales sing on their breeding ground, they string together what the biologists Roger Payne and Katie Payne described as subphrases into phrases, phrases into themes, themes into songs, and songs into song sessions. These virtual arias may last up to ten hours, making Wagner's *Parsifal* or a Grateful Dead jam session seem like relatively short performances. Within a breeding season, there are continuous changes in the structure of these elements, changes that are followed by all members of the group, providing clear evidence that individuals are imitating one or more creative singers.

Research by the Paynes and their colleagues shows that whale song consists of a nonrandom assortment of notes. There appear to be a set of formalities or conventions for how notes are arranged and then change over time. Nonetheless, the rules underlying whale song differ from those underlying human sentences. All whales in a group produce the same song at a given point in time; no whale appears to have the creative energy to stick out and sing a unique tune. Rather, it's as if all whales are members of a gospel group, singing in harmony. Although human children may mimic a parent's sentence, once they have acquired some of the formal machinery for sentence production, they are no longer restricted to this sentence. Furthermore, the notes, phrases, and themes carry no particular meaning on their own, and the same can be said of the song variants that emerge over the course of a breeding season as well as between breeding seasons. Song variants may serve an important function in identifying members of a group, and individuals capable of singing for long periods of time may be more attractive than those less able. This is not, however, the *stuff* of human sentences, that which makes the variation in word order meaningful for human language.[18]

Like whales, birds also sing. The ordering of elements—notes, syllables, phrases—in the song have been described as the songbird's syntax or grammar. The rules that underlie song construction are interesting in terms of constraints on song morphology, but are not at all similar to the grammatical rules of language. When songbirds rearrange the notes and syllables in a song, the general information conveyed is the same. With song, birds announce that they are defending a territory and looking for a mate. The order in which elements are produced also indicates their identity at the individual, population, and species level. That is a lot of information, but it is highly limited.[19]

To date, the most detailed analysis of syntactical rules in birds comes from work on the black-capped chickadee. While flying to a perch, alighting on one, or moving about, individuals produce a four-note call. The four notes, referred to as A, B, C, and D, are always produced in that sequence. Variability, however, comes when notes are repeated or omitted in a sequence, with the consequence that some calls are long and some are short. Furthermore, all calls tend to end with at least one D note.

Analyzing a large library of calls, the zoologists Jack Hailman, Millicent Ficken, and Robert Ficken found that the two most common sequences were AD and BCD, with the option of repeating one or more note types in a sequence, such as AADD and BCCDD. Some sequences never occurred or did so only rarely. For example, the D note was never followed by other notes, and the C note was never followed by the sequence BA. This suggests that the brain of a chickadee limits the way these four discrete note types are arranged. Within the limits of these constraints, chickadees have an open-ended combinatorial system, allowing them to generate a virtually limitless number of call types from the repetition or omission of notes. At a structural level, then—one that completely sidesteps the problem of meaningful variation—the chickadee call system appears to capture one aspect of natural language syntax.[20]

If the motivation for our cross-species comparison were simply structural, we might be satisfied with the chickadee results and conclude that animals other than humans have evolved rules for organizing discrete sounds into novel sequences. In thinking about the evolution of communication, however, we are forced to go beyond the structural level and ask what such rule-based systems are for, what they accomplish for the individual attempting to engage in a social interaction designed to manipulate or manage another's behavior. In this sense, the chickadee results, together with those derived from whales, songbirds, and gibbons, provide little help. For example, although the chickadee has a system for generating variable call types, we lack an understanding of the information conveyed, both in terms of what the sender apparently intends to convey and what listeners apparently extract from the call. What makes AD different from AADD? Do the extra repetitions indicate increased arousal in a particular context that is indicated by the sequence AD—flying away from a predator? Do such repetitions change the information conveyed from "flying away from a predator" to "flying toward a predator," something that a chickadee might wish to convey as it dodges an attack from a hawk or attempts to recruit help in mobbing the hawk? At present, there is no evidence that chickadees convey such meaningful information. This of course does not mean that chickadees lack this capacity.

If animals lack a combinatorial system for creating new, meaningful expressions, then perhaps Premack is right. Perhaps animals don't have anything interesting to say. Or perhaps they have interesting things to say, but the range of topics is limited, thereby alleviating the pressure necessary to evolve a combinatorial system. To examine these possibilities, we should look outside the context of song to vocalizations used in more intricate social interactions that might depend on a combinatorial system. For example, one could imagine translating certain social encounters into a form of propositional logic, one that taps the abstract concepts of subject, agency, action, cause, and consequence. Prior to engaging another individual in an aggressive coalition, the equivalent of hockey's two on one power play, individuals might run through a set of IF-THEN statements. IF I join up with Bobby, THEN our chances of beating Phil are higher than IF I fight alone or IF I join up with Wayne. And if the social environment fuels such propositional logic, then perhaps the monkeys and apes—especially those living in large social groups—are more appropriate targets of study.

Three studies, all of New World monkeys, are relevant to the idea that social pressures are related to the evolution of syntax. Primates are, after all, highly social species, with acoustically variable vocal repertoires. Researchers working on titi monkeys, capuchins, and cotton-top tamarins have concluded that these species organize their vocal repertoires around a set of rules. By using rules to combine calls, individuals can produce a greater range of meaningful expressions. Research on capuchin monkeys living in the rain forests of Venezuela provides an illustrative example.

The biologist John Robinson used field notes and recordings to show that each call type in the capuchin's vocal repertoire can be placed along a continuum from submissive to aggressive, and from contact-seeking to contact-avoiding. For example, female capuchins produce "squaws" in friendly, contact-seeking situations such as nursing and grooming. Other call types are produced in a wider range of situations. Thus, capuchins produce "trills" in neutral or friendly contexts such as playing or walking away from an infant, but also produce these calls when attacked or threatened by a more dominant animal. This classification shows that when a capuchin produces a single call type, listeners can work out the caller's approximate emotional and motivational state, but certainly cannot work out the precise context.

Robinson then explored the function of compound calls, vocalizations formed when two or more discrete call types are sequenced into a single utterance. Compound calls are formed on the basis of fairly loose rules that limit which call types follow or precede which other call types. Capuchins do not produce long strings of calls, and there is no evidence that particular call types are produced in a certain order because of their role in the sequence. In other words, there is no evidence that some calls are produced first because they function like the subject in an English sentence, whereas other calls are produced elsewhere in the string because they modify the meaning of subsequent call types. In fact, capuchins appear to create combination calls in order to express a motivational or emotional state that averages the constituent elements. The capuchin system is not, therefore, based on a set of discrete elements that can be recombined to create new, meaningful expressions. Rather, it is more like mixing different colors of paint, a blending system that averages out the contribution of each element. Thus, unlike the chickadee case, here there is no evidence of a formal structural rule to account for the sequences of call types in capuchin vocal communication. More important, perhaps, there is no evidence that such sequences add substantially to the range of topics that capuchins can talk about, and the same is true of titi monkeys and cotton-top tamarins.[21]

What to conclude? From a structural perspective, animals clearly have rules that they use to combine sound sequences. From a communicative perspective, they do not seem to have rules for recombining calls in order to generate new referential content. In the absence of such combinatorial possibilities, their vocal utterances are severely limited with respect to the range of possible meanings. Is it possible, however, that a combinatorial system might lie dormant in the brains of animals, waiting to be extracted under the appropriate conditions?

SCHOOLMARMS FOR HIRE

One of the great joys of teaching comes from seeing another individual experience the excitement of understanding a new subject that for the teacher is a source of passionate interest. When we teach, we target our efforts toward those who have the potential to learn. In the same way that it would make little sense to try and teach calculus to a zucchini, it makes equally little sense to teach calculus to a human infant. Neither the zucchini nor the infant has the potential to learn calculus, although one day the infant will acquire such potential.

The work reviewed in the last two sections suggests that the natural communication of animals lacks the referential power of most human words. Furthermore, in situations where strings of calls are arranged on the basis of rules, there is no evidence that such rules provide the basis for generating a limitless variety of meaningful expressions. It appears, then, that animals may be like zucchinis, lacking the mental tools for reference and syntax. But perhaps they are more like human infants, designed with brains that have the mental tools for reference and syntax, but lacking the appropriate environmental triggers. Perhaps animals live in a world that is more akin to the tragic cases of Kaspar Hauser and Genie, so-to-speak feral children growing up without the privilege of listening to human language.[22] If only we could reverse Dolittle's trick and provide animals with a linguistic tutorial.

In the early to mid-twentieth century, several research teams tried to teach apes how to speak. The experimenters hoped to assess whether animals such as the chimpanzee, gorilla, and orangutan could acquire spoken language. Although the question was reasonably well thought out, the experimental logic was not. Apes lack the vocal machinery required to produce most of the sounds associated with the world's languages. In this domain, then, chimpanzees are like zucchinis, or like infants up to the age of approximately four months, who have yet to develop an adult vocal tract; before human infants begin to babble, their vocal tract is primate-like, exquisitely designed for digesting food, but incapable of generating the distinctive sounds of human speech. Apes, and other animals as well, therefore lack the capacity to produce spoken language though they clearly have the potential to produce other sounds.[23] Approximately twenty years after such early failures, a subsequent generation of researchers realized that the essential question was still valid, but that they needed to explore other routes into the problem. In particular, a group of comparative psychologists and ethologists started to explore the capacity of apes and dolphins to acquire either American Sign Language (ASL) or an artificial language that preserved the essential structure and function of spoken and signed language. Here, the goal was to assess the capacity not only to produce linguistic expressions, but to understand them as well.[24]

The language studies started with the problem of reference, attempting to assess whether animals were capable of understanding that symbols stand for objects, actions, and abstract concepts, that they can refer to things that are out of view, can be used communicatively to make requests, describe situations, and express ideas, and consist of elements that can be recombined to form new symbols. The earliest stages of the tutorial involved training individuals to learn the association between a symbol and a referent. The experimenters then tested their subjects' capacity to generalize to novel contexts and instances of the symbol's referential class. If they acquired an understanding of reference, then having learned the sign for "apple" while tested in the presence of one trainer, they should be able to apply this sign to an array of apple types, representations (photographs, real apples, drawings), and situations (other trainers, animals). In some studies, the apes were trained in an environment designed to simulate a child's experience growing up in a home. In other cases, the apes were reared in the highly controlled context of a laboratory, a pool for Lou Herman's dolphins Ake and Phoenix, a cage for Herb Terrace's chimpanzee Nim Chimpsky. Those using sign language had to mold the shape of their subjects' hands since none of the animals showed strong evidence of spontaneous imitation (see chapter 6). In both the signing and artificial language research programs, a subject's success in using or responding to a symbol was met with food reinforcement and often verbal praise.

Over a period of several years, all the research teams reported success of some sort; animals appeared to have acquired hundreds of symbols. For example, the Gardners' chimpanzee Washoe originally learned the sign for "open" when confronted with a closed door. He then generalized to other contexts, using the same sign when confronted with significantly different but closed things, such as containers and even faucets. The challenge comes, however, in showing that the gesture is not the result of conditioning—associating through reinforcement an action or object with a sign. Rather, it is necessary to show that the chimpanzee's use of or response to a sign is like our command of words, neither inevitable, highly specific, nor restricted to a particular context. More simply stated: when an animal responds to a symbolic gesture or produces one, is it doing anything more than a dog that moves to the door when his owner takes the leash out? In this case, the dog has certainly learned how to create an opportunity to go outside, but such know-how is fundamentally different from the awareness required to infer that a particular signal causes a change in another's beliefs.

To address the question of similarity, we can compare the use and comprehension of symbols by animals with that of developing human children and that of adults, breaking the comparison down into three separate issues. First, do animals grasp the idea that word-like symbols can refer to physical objects, events, and abstract concepts, to tokens such as "banana," "eat," and "different"? Observations by several research teams suggest that they do, particularly in the context of tasks that bring about specific responses, such as receiving a food reward. Thus, when David Premack presented his chimpanzee Sarah with half an apple and a glass that was filled halfway with water, Sarah produced the symbol for "same." This provides evidence that at least some animals can learn a symbol's referent, in the sense that it stands for something, even something abstract. Additionally, many of the observed errors in symbol use or comprehension appear to be cases of generalization, thinking that a symbol stands for a broader class of referents than it actually does. Thus, like human children who say "doggie" for all four-legged creatures, the Gardners' chimpanzee Washoe produced the sign "flower" for flowers, as well as other objects and locations with strong smells.

Second, is the animal's use of a symbol completely predictable from the context experienced or do animals use symbols to refer to objects that are out of sight, events that have yet to occur, or actions that they are planning? Unfortunately, most of the experiments were set up to elicit a symbolic response to the presentation of an object or event, thereby precluding spontaneous, unpredictable commentary. There are, however, some reports of spontaneous commentary, primarily in the form of requests for food or access to a favorite trainer, answers to the presence or absence of an object, and a few cases where novel sign combinations were produced for an object lacking an explicit sign. In Herman's work, for example, if a dolphin was asked "Hoop question"—simulating a query about the presence of a hoop— the dolphin returned a "no" response if the hoop was absent from the pool. This shows that the capacity to sign is not restricted to the physical presence of an object. In terms of novel signs, Penny Patterson reports that the gorilla Koko signed "white tiger" when she was presented with a toy zebra, an object for which she lacked a sign. This shows some capacity to invent new sign combinations from preexisting signs. Far more impressive along these lines are results from Sue Savage-Rumbaugh's studies of the bonobo Kanzi and his use of a symbolic keyboard. When he heads out on his own into the forested enclosure, Kanzi often takes his keyboard. During his travels, he apparently *talks to himself*, using the symbols to comment on what he sees or to reveal his travel plans, including an indication of where he is going and what he intends to do. Furthermore, and in contrast to the other language-trained animals, Kanzi's symbol use is less closely tied to requests from his trainers. For Kanzi, the symbols are not only instruments for action, but meaningful representations of things in the world. Still, most of Kanzi's symbolic expressions are requests, and in this sense, he is like the young child with a "me, me" attitude.

As might be expected, there have been quite acrimonious debates over the quality of these data, both among the primary researchers and from those looking on from the outside. Furthermore, experiments by the psychologists Ronald Schusterman and Robert Gisiner argue that sea lions acquire the same kinds of referential capacities as dolphins and apes. This suggests, contrary to the view held by many in this field, that you don't really need a big brain to acquire at least some linguistic capacities; relative to dolphins, gorillas, and chimpanzees, sea lions have quite tiny brains. Nonetheless, I think it is fair to conclude that given extensive training, at least some animals have the capacity to go beyond what has been shown using natural vocal signals. Some animals can acquire an understanding of reference that is independent of the stimulus, and are perhaps even capable of generating novel expressions.

A final issue concerns whether or not language-trained animals understand that symbols, such as the signed gestures that constitute ASL, consist of subunits that can be recombined to create new signs. For example, the primary units in an ASL sign are hand shape, hand motion, and place of articulation (where the hand is placed relative to other parts of the body). By recombining these components, one creates new signs. The Gardners and their colleagues observed that chimpanzees were most likely to make signing errors when two or more signs shared the same place of articulation. This pattern suggested that the chimpanzees were attending to the particular features underlying each sign. Such errors could, however, be due to either perceptual mistakes, such as confusing two different signs with the same hand shape but different motion, or a relatively poor understanding of a sign's specific referent. For example, the chimpanzees may think that a particular sign refers to a greater range of objects than it in fact does. Moreover, even if the error is perceptual, we still haven't learned how chimpanzees *read* the signs of ASL, whether they understand that they can generate new signs with new referents by recombining components of a sign.

A peculiar twist to the entire story about language-trained animals emerged in the late 1980s when Savage-Rumbaugh and her colleagues reported that Kanzi had acquired, in the absence of explicit training, comprehension of spoken English. At first blush, it is not even clear what this claim means. We might say that dogs understand spoken English, and so do animals involved in circus acts, jumping through a ring of fire at the command of their trainer. But this is not the kind of comprehension that was being claimed for Kanzi, and more recently, some of the other bonobos that have been reared under comparable conditions. When we call Fido the dog, the name "Fido" certainly grabs his attention, but so will "Ido," "Wido," and "Bido." Fido has learned to pick up on certain sounds because they are associated with certain kinds of events—going for a walk, eating, and playing fetch. But Fido has no understanding of words in the sense of sounds carrying distinctive acoustic features and referents. Kanzi, in contrast, couldn't be fooled by the "Fido-Ido-Wido-Bido" trick. In responding to a speaker, he was listening and responding to those sound contrasts that allow us to distinguish words that sound alike but mean different things because of small changes in the initial or terminal syllable. Moreover, Kanzi appeared to be listening to the precise order of words in a sentence and using order to compute the appropriate action. Savage-Rumbaugh claimed that Kanzi had learned—and continues to learn—the correct referents to several hundred spoken words, and learned that the meaning of an utterance depends on syntax, the order of words, and the rules that allow some words to occupy a particular position in a sequence.

Savage-Rumbaugh's claim was radical on two counts. First, in no other study, including those on the closely related chimpanzee, had an animal demonstrated any facility in learning spoken language. In fact, even Kanzi's mother, Matata, failed to learn spoken language. Because Matata started training as a young adult, whereas Kanzi was born into the lab, Matata may have missed the critical period for exposure to human language. As we have learned from the few cases of feral children, insufficient exposure to language before puberty leads to either no linguistic capacity or a highly impoverished one. Perhaps there is a critical period for apes to learn some aspects of spoken language. Second, all previous attempts to show that animals can produce a combination of ASL signs with a syntactic structure, or respond appropriately to ones produced by a human trainer, have generally failed.[25] How, then, was Kanzi doing it, and what, precisely, was he doing?

If Kanzi understands the importance of word order in the way that we do, if he has truly acquired some of the crucial rules guiding natural language syntax, then certain conclusions follow. For example, if he understands the meaning of a word in one context, he should understand it in other contexts as well, assuming the word is embedded in a sentence that has words that he also comprehends. In some tests of sentence comprehension, this appears to be the case. Thus, for example, although Kanzi was extremely possessive of his toy balls, when Savage-Rumbaugh asked the novel question, "Can you throw your ball in the river?" he immediately went over and did so. This sentence could, of course, be processed the way the dog did it in one of Gary Larson's cartoons, where the owner speaks and the dog hears "Ginger blah blah blah blah Ginger blah blah, etc." Kanzi might just hear "Blah blah *throw* blah *ball* blah blah *river*." Since the "river" can't be put into the "ball," the appropriate reading of this sentence is limited. To meet this criticism, Savage-Rumbaugh and her colleagues ran other tests, ones where a mistake would be obvious. Thus, for example, they asked, "Can you throw a potato at the turtle?" With such questions, if Kanzi only heard "Blah blah *throw* blah *potato* blah blah *turtle*," he might throw the potato at the turtle or the turtle at the potato. He rarely made such mistakes. In other domains, however, Kanzi made significant errors. When asked to do something with two or more objects, he typically operated on only a single object. Thus, when asked, "Give Sue the hat and potato," he would hand over the hat, but not the potato. At some level, then, Kanzi appears to be attending to the order of certain key words, and more recent tests with other young bonobos suggest similar findings. Furthermore, these animals appear to comprehend that words can be classified according to quite abstract categories such as actions, objects, and locations. With rules for ordering words and an understanding that words fall into abstract categories, these bonobos have acquired a system of comprehension that is highly expressive, involving what many linguists would consider the proto-syntactic structures of action-object or action-object-location.

Much of the scientific community has been unwilling to accept these new and highly controversial findings. Some of the criticisms, I believe, are completely misplaced, while others will require more thought and more data to resolve. Thus, some argue that structure such as action-object-location, though somewhat abstract, is nothing like our concepts of noun, verb, subject, predicate, and so forth. Without such mental tools, the bonobo's capacity for language is relatively limited on the expressiveness scale. Although the distinction between the kinds of syntactical structures Kanzi has acquired and our own natural language syntax is important, this critique misses the point because it does not consider the evolutionary problem. In particular, if we are interested in how natural language syntax evolved, we need to entertain the possibility of a different system of precursor rules, ones that may have provided the foundation for subsequent evolution. I see no reason a priori why natural language syntax requires one and only one rule-based system. It seems theoretically possible that an earlier system depended on different kinds of rules.[26] If this account turns out to be right, there are some interesting ideas waiting around the corner. In particular, linguists, evolutionary biologists, and anthropologists will need to link arms with computer scientists and attempt to provide an empirically grounded simulation of how rule-based systems evolve, how one might go from an action-object rule to something more powerful, more abstract.

What is debatable, and critical to considering human language evolution, is whether the rules acquired by Kanzi and the other bonobos enable them to generate new and meaningful expressions. How significantly limited are they in conveying their thoughts, plans, and emotions? And irrespective of how this problem is sorted out, why don't bonobos use their capacity for reference and syntax to communicate with each other naturally? Perhaps we haven't looked hard enough? Perhaps, but I am skeptical. Even Savage-Rumbaugh, who has watched bonobos at close range and described many of their vocalizations, has failed to come up with a single example. Similarly, Frans de Waal's analysis of the bonobo's repertoire reveals nothing like a referential signal or a system of rules for putting together novel sequences of sounds.

JAILED THOUGHTS

"As Gregor Samsa awoke one morning from uneasy dreams he found himself transformed in his bed into a gigantic insect." So begins Franz Kafka's haunting tale, "The Metamorphosis." In an episode of *Star Trek*, a peculiar force invades the ship, causing the crew members to revert back to an ancestral animal form—they have devolved. Both scenarios capture something profound about human nature, about what it would mean to lose those characteristics that we consider precious and unique. Kafka's scenario is by far the more ominous one because the transformation targets Gregor's body, leaving his mind and his emotions in a pristine state. Trapped in a hard shell, he experiences the world in the same way that other humans do, but he lacks a voice to express what he experiences. He is a prisoner, silenced by the body of an insect.

If my thoughts on animal communication are correct, then a treasure trove of evolutionary problems emerge. Most animals are like the unfortunate Gregor Samsa after metamorphosis. They are Kafka-creatures, organisms with rich thoughts and emotions but no system for translating what they think into something that they can express to others. By making this claim I do not mean that animals, lacking a human language, have the kinds of thoughts that Samsa-as-beetle has. Without a doubt, our thoughts are different, and language has contributed to this difference, though I am not convinced we know exactly how. I am also not claiming that animals lack the machinery to express certain aspects of their feelings and thoughts. Rather, they are severely handicapped, especially when it comes to communicating about details of their social relationships. In fact, the studies by Sue Savage-Rumbaugh and Mike Tomasello suggest that even though apes reared in a human environment acquire cognitive abilities that naturally reared apes do not, the human environment is an insufficient condition for the acquisition of a natural language. Without a sufficiently expressive system, the rules for societal organization will be myopically focused on the present, never forward looking, assessing how right and wrong might change under certain conditions. There can never be proscriptive recommendations for how things should be. And there can never be hypotheticals, counterfactuals, or discussions of possible outcomes, possible wrongdoings, and possible gains. This, in turn, leads to a second puzzle, one that places the evolution of humans in a new comparative perspective. The earliest humans were endowed with a formidable mental life, but had to wait patiently for the arrival of a new tool, the gift of language. When this mental tool arrived is a mystery, but its birth emancipated a mind that was humming along with its own private thoughts. This emancipation led to a society of individuals that not only knew how to cooperate and cheat, but could explain why someone may be put up on a pedestal while others would be punished.

The moral sense represents a balancing act. Each player must inhibit the selfish tendencies that are endowed by Nature in order to maintain societal conventions.

9

Moral Instincts

An elephant stands still, using her trunk to hold the bones of a dead family member. A lion, upon taking over a pride, immediately kills all the newborn cubs and then begins a mating sequence involving thousands of copulations. Having been attacked by the dominant male in his group, a subordinate male stump-tailed macaque walks over, grimaces, and then receives a hug from the benevolent leader. A male duck aggressively mounts and copulates with a smaller, apparently unwilling female, who attempts to flee. To obtain a stash of meat sitting out of reach on a platform, two hyenas simultaneously grab hold of some dangling rope. Pulling in perfect rhythm, they release a latch that delivers the food.

These observations from scientists' notebooks describe events with little interpretative spin. All share a number of similarities and pose several distinctive problems. From our own perspective, it looks as though the animal is acting as a function of a specific emotion and a specific goal, such as *sorrow* from the loss of a family member, *cooperation* designed to achieve a common goal, *respect* for the rules of property, *love* between mothers and their young, and *fear* of attack from a more powerful group member. Are these appropriate descriptive terms and interpretations? The bond between mothers and their young is unquestionably strong and certainly based on an instinct. Yet how can we know what the mother feels when her infant nurses or is threatened by a baby-killer, if she feels anything at all? Nonhuman primate mothers often show off their young to others, as if they feel pride.

However, they almost never look into their infant's eyes and make adoring noises as would a human mother with her child. A male duck forces himself sexually on a female in an act that would be considered rape if it occurred between a man and a woman. Yet how can we know what the female duck feels? Does she feel wronged? Does she have a sense that a social rule has been violated? Or does she feel that this is just another day in the life of a female duck? When hyenas work as a team to obtain meat, do they expect an equal division of the spoils? Do they have a sense of fairness? Do they enter the hunt with a mutual agreement, an understanding that each hyena has a role to play and is responsible for a specific result?

To answer these questions, I take an admittedly reductionistic approach. Following in the footsteps of the philosopher Immanuel Kant, who drew a comparison between moral theory and chemistry in the conclusion of his critique of reason, I look at morality the way a chemist looks at the structure of a crystal and the factors that preserve its integrity. Instead of asking whether animals are moral, I ask such questions as, Do animals have a sense of fairness? Do they understand the distinction between right and wrong? Do they place values on certain kinds of interaction and generate expectations based on a set of norms? Do they empathize, feel guilt or shame? Can they inhibit actions that are likely to cause harm to others? Do they punish those who violate social rules? These questions are not easy to answer. They do, however, force us to focus on significant components of morality, components that ultimately guide moral action. This provides a richer understanding of how the moral mind evolved, and the extent to which other animals share in a moral fabric.[1]

SERPENT SAYS

Once upon a time, a man named Adam lived all alone in the Garden of Eden. God created Eve in order to reduce Adam's loneliness, and Adam was satisfied. Life is idyllic for the couple until the serpent shows up. As the biblical scholar Elaine Pagels points out in *Adam, Eve, and the Serpent*, there are several theories about the serpent, what he represents, what his goals are, and what we should infer from his actions.[2] Whether the serpent represents wisdom, Christ in disguise, or God's messenger, the consequences of his interaction with Eve are the same. By tempting her to eat from the tree, he opens up her eyes to the difference between good and evil, to value-laden judgments and ethically profound concerns. The Lord punishes Adam, Eve, and the serpent for their actions, thereby establishing a set of norms, as well as the potential consequences of ethically relevant actions.

Leaving to one side the biblical interpretation, where do our ideas about right and wrong, good and evil, or selfishness and altruism come from? Although "right" and "wrong" are abstract labels, they may derive from concrete roots grounded in emotion. We feel guilt when we violate a rule, shame when we fail to help a friend in need, and gratification when we give to a charity. Most if not all of these feelings are tied to a range of tolerable behaviors given the circumstances, what the philosopher Allan Gibbard has described as normative emotions.[3] Although animals cannot verbally label actions right or wrong, do they understand why certain actions violate rules and do they have the relevant emotions?

Emotions are *about* objects and events from the past, present, and future. These objects and events can occur in either the external environment or the inner workings of the mind. To understand emotions, then, we need to assess how animal minds represent particular objects and events in the world. That is, we need to return to some of the mental tools discussed in the previous chapters and assess how and why some objects come to be associated with or elicit particular kinds of emotions, thoughts, and intentional actions.

Once again, the study of human infants provides key insights. Infants ultimately develop into beings that participate in a moral society, equipped with the moral emotions such as shame, guilt, embarrassment, sympathy, empathy, loyalty, and fairness. What conceptual tools, then, do infants bring to the task of moral action? Infants smile at some things and frown at others. Such expressions emerge early in life and represent categorical decisions about the kinds of things that are enjoyable and rewarding on the one hand, and sad or frightening on the other. Inanimate objects on their own tend not to elicit such expressions. Rather, whether or not an infant smiles or frowns at a ball or rock depends on the kinds of experience the infant has had with that object. As several authors have demonstrated, a parent's response to something or someone can determine how the infant responds in the future. By means of pointing, staring, and making faces, parents impose a heavy hand on the infant's emotional responses.[4] Such experience is not, however, the final word. Children appear to be born with biases about objects, biases that allow them to generate expectations about what some objects can or should do. The animate-inanimate distinction may well be the most important bias of all.

Before children acquire a full-fledged theory of mind, they appear to build a theory of moving objects. This theory of moving objects has at its core one of the universal tools—knowledge of objects. In particular, before infants reach their first birthday they generate expectations about objects based on how they move and what causes them to move. By six months, human infants expect hands to be capable of both lifting and supporting a ball, but do not extend this expectation to physical objects such as sticks. Experiments by the developmental psychologist Alan Leslie show that six-month-old infants also understand that self-propelled objects have different capacities for moving in an environment than do objects that require an external force to move. For example, six-month-olds expect a stationary blue box to move if it is contacted by a moving red box, but are surprised to see it move in the absence of contact from an external force. By twelve months, infants attribute goals to self-propelled objects. Thus, infants expect a ball to move in a straight line toward a goal if there are no barriers in the way, but expect the same ball to take a more roundabout path if something is in the way. Clearly then, motion provides important information about the kind of object that is moving, and infants are sensitive to motion at an early age.[5]

Infants are sensitive to motion, but do they assign emotions or motivations to objects based on *how* they move or interact with other objects? A series of experiments in the 1940s and 1950s provide some answers. The experimental psychologists Heider and Simmel showed adults a short animation on a monitor. In the initial display, a small circle and triangle are located inside a large square. Next, the circle moves around the screen, trailed by the triangle. Eventually the circle ends up in a corner of the square, and the triangle, with one of its points leading, makes brief, repeated moves toward and away from the circle. Adults tend to interpret this sequence by saying that the triangle *chased*, *cornered*, and then *attacked* the circle.[6] Though this experiment elicits some overinterpretation, it reveals the power of motion to trigger emotions, at least in adults. Do they trigger comparable emotions in infants or other animals?

David Premack and Ann Premack provide an ingenious solution, using a series of computer animations to assess whether one-year-old human infants attribute specific emotions to objects that move in particular ways. The Premacks started with four assumptions. First, living things move on their own, whereas nonliving, physical objects can move only if contacted by something else. Second, psychological objects have goals, and move in such a way as to achieve their goals. Third, when an object moves on its own and is goal-directed, the viewer will automatically attribute thoughts and emotions to the object. This perspective enables observers to make inferences about the object's intended actions.[7] Fourth, when two or more objects with the properties listed above interact, viewers will assign an emotional label to the interaction. For example, in the Heider and Simmel animation, adults interpret the triangle's contact with the circle as negative.

The Premacks created four animations with a black and a gray circle, two involving positive emotions (caress, help) and two involving negative emotions (hit, hinder). In the *caress* sequence, contact between the circles was gentle, and neither circle was deformed upon contact. The parallel negative sequence, *hit*, involved forceful contact, with the gray circle deformed by the black circle. For the *help* sequence, the black circle dropped down and gently pushed the gray circle up and through an opening in a barrier. In the contrasting sequence, *hinder*, the black circle blocked the gray circle's passage through the opening in the barrier. To determine what infants think and feel about these animations, Premack and Premack used a habituation procedure. The logic was simple: repeatedly play back one sequence until the infant is bored and looks away, then present one of the remaining three sequences. Infants should be more interested in seeing a sequence from a new emotional category than a sequence from the same category if they see these animations as carrying meaningful information about emotions. For example, after watching the black circle repeatedly caress the gray circle, a positive emotion, they should be more interested to see the black circle hinder the gray circle, a negative emotion, than help it, another positive emotion. This logic should apply to all possible combinations that contrast positive and negative emotion. Conversely, when two emotions from the same category are contrasted (e.g., *caress* and *help*), infants should continue to be bored.

The Premacks' findings supported these predictions. Once an infant habituated to one sequence, her interest in the second depended on the emotional quality or valence of the animation. Having seen several reruns of the caressing animation, infants looked longer in response to the hitting animation than to the helping animation. Having seen hindering over and over again, they looked longer at helping than at hitting.

Do these results prove that infants assign emotions to the animated sequences? Although suggestive, these experiments run into the same difficulties we confronted in discussing the universal toolkit, particularly representations of object and number. Specifically, what kinds of inferences can we make about a subject's thoughts or emotions from studies that use looking time as a measure? Infants certainly perceive a difference between animations. The data *say* so. But do they experience kindness when they see the black circle move to help the gray circle through the opening? Do they empathize with the gray circle when its attempt to pass through the opening is blocked by the black circle? If the black circle is perceived as negative, would the infant prefer to interact with the gray circle if given a choice? Are objects that interact positively preferred over objects that interact negatively? If the circles came to life as 3-D objects, would infants rather cuddle with black fuzzy spheres than gray fuzzy spheres? To address these questions, we can combine measures of looking time with studies that allow infants to play with objects, as well as physiological measures such as heart rate that may allow us to assess what emotions infants experience when they view different sequences. We might expect, for instance, that the physiology associated with seeing *caress* would be comparable to that associated with seeing *help*, and significantly different from that associated with seeing *hit* and *hinder*. Given the increasing sophistication of research on the neurobiology of emotion, such expectations are more than wishful thinking.[8]

What about animals? Most, if not all, animals act as if they classify the world into animate versus inanimate objects, perceive animate objects as having goals, and assign particular emotions to animate objects as a function of their actions. However, if we are going to evaluate the possibility that animals have a moral sense, then we must say more than the "as if" phrase captures. For example, do animals make a distinction between objects that move on their own and objects that require an external force to move? And if so, do they attribute different emotions to objects based on how they interact with other objects?

Consider the following looking-time experiment that my students and I ran with cotton-top tamarins. We exposed tamarins to a box with two identical compartments, separated by a partition; the partition was opaque, with a hole that allowed objects to move between compartments. We first familiarized subjects with all the objects and events to be used in the critical test trials. These included a frog, a mouse, a stationary cluster of Froot Loops, a ball rolled by an experimenter, a self-propelled clay face, and a toy tamarin that jiggled in place. During familiarization, none of the objects moved through the partition, and the tamarins never saw anything unexpected. No physical violations, no magic. By the end of these trials, all objects were equally boring.

The test trials were designed to ask tamarins this question: if an object is placed into one chamber and then hidden by a screen, where should the object be following the screen's removal? Human adults expect the object to remain in the original chamber if it can't move on its own, and expect self-propelled objects either to remain in the original chamber or to move to the other side. The tamarins apparently generated the same expectations, with one important exception. Although they looked longer when the Froot Loop cluster appeared in the opposite chamber from its starting point, they also looked longer when the other inanimate objects changed sides. This is surprising given that the clay face was self-propelled.

If looking time provides a measure of what an animal expects, then tamarins do not expect inanimate objects to move to a new location, even if they are self-propelled. Thus, although the toy tamarin and clay face moved on their own, the tamarins apparently expected them to stay in the same chamber. These results suggest that tamarins form different expectations for objects that are alive than they do for objects that are not.[9]

The tamarin experiments, like those conducted by the Premacks on human infants, are open to alternative explanations. In particular, given the small number of objects presented, we still don't understand what tamarins expect from animate as opposed to inanimate objects, nor do we fully understand how they use motion to make predictions about an object's behavior. How would the tamarins respond to other animate creatures with strikingly different patterns of locomotion, creatures such as snakes or birds? What kinds of expectations would they generate for a dead mouse, or one that had been anesthetized and thus fails to move? Although the clay face, toy tamarin, mouse, and frog move on their own, they may differ in more subtle aspects of motion; living creatures move in ways that are different from nonliving creatures, and tamarins may be sensitive to this distinction. Additional experiments and observations will be necessary before we can conclude that tamarins generate expectations about object location according to whether they are alive or not. These experiments will help us understand whether animals, like tamarins, can use the mental tools of object, number, and space to build a more sophisticated set of expectations about objects with minds.

If we assume that tamarins and other animals draw a distinction between animate and inanimate objects and use motion as a discriminating feature, is there any evidence that they use a combination of motion cues together with goal-directed action to attribute emotions to a group of interactants? The Premacks' experiment has not been run with animals yet, but there are experiments that address the kinds of emotions animals attribute to other individuals. The emotions used in these experiments are critical elements of morality.

In 1996 a visitor to the Brookfield Zoo in Chicago captured an extraordinary event on his video camera. A young boy fell into a gorilla enclosure and was knocked unconscious. Within minutes, a female gorilla named Binty approached, picked the unconscious boy up in her arms, and cradled him in her lap. She then walked over and gently put the boy down in front of the caretaker's door.

This event captured the attention of the national media, with headlines running. "GORILLA SAVES BOY," "ALTRUISTIC GORILLA," and "THE MORAL SENSE." Most reports suggested that Binty saved the boy, acting altruistically and empathetically. Although there is no ambiguity about *what* Binty did, there are many possible interpretations of her thoughts and emotions. Did Binty realize that the boy was unconscious? Would she have acted in the same way to other novel objects, a conscious boy, a cat, a teddy bear, or a bag of potato chips? What would her response have been if the boy suddenly woke up in her arms? There are, unfortunately, no answers to these questions.

The most intriguing aspect of Binty's behavior, of course, is the possibility that she acted altruistically. Not the selfish-gene brand of altruism that sociobiologists have discussed, but the kind of altruism that is typically associated with the likes of Gandhi and Mother Teresa, individuals who devoted their lives to the well-being of others. Did Binty act out of the goodness of her heart, showing kindness to another creature, investing herself in its well-being and ultimate happiness? Did she understand the consequences of her actions? Did she think that she was doing the right thing? Did she feel, as ethically sensitive humans often do, that picking up the boy was the only *possible* choice, that there was really no other appropriate response? Did she not only feel sympathy for the unconscious child, but empathy for what it might be like to be in such a state? We can't answer these questions based on one observation, even when we are watching humans. Fortunately, there are several experiments and observations that begin to provide the necessary insights.

The following experiments were conducted in the 1940s and 1950s and were designed to enrich our understanding of animal emotion and thought. When they were conducted, the standards for animal welfare were minimal. By today's standards, some if not all of the experiments would be considered unethical. The reason for discussing them here is that they provide us with a potential window on animal emotion, a window that has yet to be opened by more recent experiments and observations.[10] The fact that I discuss this work should not, however, be read as an endorsement. It most definitely is not.

A crucial component of altruism is that someone performs an act for the well-being or good of another. The animal psychologist Russ Church designed an experiment to determine whether genetically unrelated rats would engage in such behavior. First, he trained a rat to press a lever for food, and then he changed the result of pressing the lever. On some trials, pressing the lever resulted in food, whereas on other trials it delivered a shock to a second rat visible in an adjacent chamber. Rats in control of the lever immediately stopped pressing the lever, thus giving up the opportunity to eat, but saving another rat from shock. After several days, some rats started pressing the lever, but they pressed less than when no shock was delivered to the rat next door.

Do rats stop pressing because they recognize the pain being inflicted on another rat? In a second experiment, an experimenter trained a control group of rats to press for food, but with no one else present and no shock. One test group of rats pressed for food, but sometimes the experimenter delivered a shock. A second test group of rats pressed for food, but sometimes received a one second period of shock, while another rat in view received a thirty second period of shock. The first test group's response rate dropped to about 20 percent of the control's, while the second test group's response rate dropped to zero. Ten days later, at the end of the experiment, the second test group responded less than either of the other groups.

These experiments suggest that rats are willing to eat less if, by inhibiting particular responses, they benefit another rat. This looks like altruism based on sympathy, perhaps even empathy. Several questions remain, however. Perhaps rats stop pressing the lever because the other rat's screaming and wriggling are annoying, physically unpleasant. If this is the case, then the rat pressing the lever did not feel sympathy, a feeling of sadness caused by another's pain. Nor did the lever-pressing rat feel what it would be like to be another rat receiving shock—empathy. Rather, the rat's screaming and wriggling are unpleasant. All animals will find ways to convert an unpleasant situation into something pleasant or at least less unpleasant. Further, the rat's capacity to inhibit pressing appears to fade over time, even though by pressing, he continues to deliver the shock. In terms of sympathy or empathy, the rat's selfishness seems to take over. The rat's reaction is understandable— wouldn't we start pressing as well if we got hungry?

To determine whether rats would reduce another's potential suffering even when they gained no direct benefits, researchers first trained one group to press a bar upon seeing a light; failure to press the bar resulted in a shock. In phase 2, pressing the bar while the light was on caused a Styrofoam block to be lowered to the ground, left there for fifteen seconds, and then hoisted back up; failure to press the bar again resulted in a shock. In phase 3, the shock was eliminated; thus the rats no longer had any motivation to press the bar. These rats were then divided into two groups. The experimenter presented group A with the light and Styrofoam block hoisted above, and group B with the light and a rat suspended in the same position as the Styrofoam block; while suspended, the rat squealed and wriggled. Whereas the rats in group A almost never pressed, the rats in group B pressed a lot, thereby lowering the suspended rat to the ground and relieving him of his distress; the rate of bar pressing among the rats in group B decreased from day one to two.

The rats in this experiment were not responsible for causing pain to another. Rather, the experimenters presented these rats with an opportunity to relieve another's pain without much cost. The pattern of their response suggests that they were acting altruistically, helping another rat in need without receiving any direct benefits. We are, however, left with the same set of questions that we raised for Church's experiment. Was the increase in bar pressing a selective response to a rat in distress? Or did the rats in group B press the bar because they couldn't stand the sound and sight of another rat screaming and wriggling? And why did the rate of bar pressing drop after day one? Perhaps these rats got used to the screaming and wriggling.

To determine whether rats bar press to reduce another's suffering (an *altruistic* act) or to reduce the unpleasantness of another's noises (a *selfish* act), an experimenter presented one group of rats with white noise, and a second group of rats with recordings of rat squeals. Pressing a bar terminated the sound playback. After several trials, the groups switched conditions, so that the second group heard white noise and the first group heard rat squeals. Both groups pressed the bar more in response to white noise; this suggests that white noise is more unpleasant than rat squeals.

These results show that white noise is annoying, and that rats will press the bar to eliminate whatever annoys them. This fact, however, does not allow us to reject the idea that pressing the bar to rat squeals—or to the sight of a squealing rat—is altruistic. Rather, this result shows that rats will press a bar to turn off a variety of unpleasant stimuli, including rat squeals.

Unlike rats, most primate species express their emotions through a rich repertoire of facial expressions.[11] The psychologist Robert Miller and his colleagues ran experiments on rhesus monkeys to assess whether they would act altruistically toward each other based on particular facial expressions of emotion. An experimenter trained a rhesus monkey to avoid shock by pulling a lever in response to hearing a sound. On the test day, one animal playing the role of the actor was placed in a room with a lever and a video image of a second animal, the receiver, located out of sight and hearing range. The receiver was presented with the sound played back during training—the one associated with shock—but without access to a response lever. The assumption underlying this experiment is that the receiver will hear the sound, anticipate shock, and respond with a facial expression, one denoting fear. If the actor reads the receiver's facial expressions, then he should use this information to time his response. If the actor fails, then both animals receive a shock. Since shock trials were presented randomly, and since neither animal could hear the other, there was no way to predict the timing of a response except by using the receiver's image in the monitor.

The actor pulled the lever significantly more when the receiver heard the sound than during silent periods; this showed that the actor was able to read the receiver's facial expressions. Miller and his colleagues suggest that both animals had acquired a cooperative response: to avoid shock, the receiver must signal some information and the actor must read her signals.

Is this a valid conclusion? Did the receiver intend to provide information to the actor? Was this a cooperative effort? I am not convinced that the experiment warrants such conclusions, although it is suggestive. The receiver must have felt helpless, distressed, and afraid. But to say that she was signaling, one would have to show that she knew about the actor's presence. Given the experimental setup, however, she most certainly did not. Rather, the receiver's response was elicited by the sound, perhaps as reflexively as our knee-jerk response is elicited by the doctor's tiny mallet. Further, and against the cooperative explanation put forward, it seems more likely that the actor picked up on a change in the receiver's activity, one that was sufficiently consistent as to provide an accurate predictor of things to come. Using an expression to predict a response is not the same as seeing the expression as an indication of another's emotions.

This experiment leaves many loose ends. Although it is clear that rhesus can avoid shock by attending to a facial expression, we don't know whether this response is motivated by empathy, an emotional capacity that is necessary for true altruism. One has to feel what it would be like to be someone else, to feel fear, pain, or joy. We don't know whether the actor was even aware of the receiver's feeling: to be perfectly blunt, there is no reason for the actor to care. From the actor's perspective, all that matters is that the image displayed on the video monitor functions as a reliable predictor of shock. A better experiment would still allow the actor to see what was happening to the receiver, but would restrict the shock to the receiver.

To push the issue further, Miller and his colleagues ran a final experiment. An actor was trained to pull one of two chains to receive food. Next, a second rhesus monkey—the receiver—was introduced to an adjacent cage and the animals were allowed to see each other. In the second phase, the experimenter changed the consequences of pulling a chain. One chain continued to deliver food, whereas the second delivered a shock to the receiver. Of the actors tested, most showed a substantial drop in the number of responses to the chain delivering the shock, especially when contrasted with the chain delivering food. One subject stopped pulling both chains for five days, and another for twelve days. This self-starvation happened more in animals who had experienced shocks themselves. When the actors were paired up with new receivers, most continued to refrain from pulling the chain delivering the shock. Interestingly, there was a tendency for pairs that knew each other well to show more altruistic behavior than pairs who were unfamiliar. It's as if altruism was affected by friendship, shifting the economics in favor of helping another in need if the other was well known.

What is most remarkable about these experiments is the observation that some rhesus monkeys refrained from eating in order to avoid injuring another individual. Perhaps the actors empathized, feeling what it would be like to be shocked, what it would be like to be the other monkey in pain. Alternatively, perhaps seeing someone shocked is unpleasant, and rhesus will do whatever they can to avoid unpleasant conditions. Although this has the superficial appearance of an empathic or sympathetic response, it actually may be selfish.

For many, the experiments described above are unethical because we should not harm animals for our own intellectual benefit. For others, such experiments are justified because of what we learn from them. There are no easy solutions here, and I will not engage this issue further. However, a growing body of work addresses comparable issues, and does so by turning the ethical issue on its head. Specifically, over the past ten years the ethologist Marian Dawkins and her colleagues have explored the needs of animals by using an economic approach to their well-being. The logic of her work is simple. Animals will work harder, and perhaps incur greater costs, for resources that they really want. As Dawkins and others have argued, from measuring what animals want, we can derive a better understanding of their needs. We can thus begin to look at the psychology of desire by exploring the economics of choice. Consider an empirical application of this logic.

In Britain, hens used for egg production are kept in relatively small cages. At one point, the government attempted to reduce costs by removing the chipped wood lining the cage floor. Animal welfare activists responded that hens use the wood to scratch, and this is part of their normal behavior. Put simply, hens *need* chipped wood. To test this idea, Dawkins created a chamber with two identical compartments, separated by a door. In one compartment, however, chipped wood lined the floor, while the other compartment was bare. When the experimenter placed the hen in the bare compartment, the hen immediately walked over to the other side. To show how much hens need chipped wood, Dawkins increased the tension on the door's hinge, making it more difficult to open. Nonetheless, the hens worked to open the door, sometimes injuring themselves simply to gain access to the chipped wood.

Dawkins's experiment provides a simple technique to understand animal emotion and motivation. Other researchers have used this technique to show that animals will work hard to obtain more space or to gain opportunities to interact socially with members of their species. Spatial confinement and social isolation are clearly negative experiences for many animals. Dawkins's experiment allows animals to show us.[12]

What should one conclude from the collection of experiments described above? I suggest that although many animals have the mental tools to distinguish between living and nonliving things, to use object motion to generate expectations about behavior, and to have emotional experiences about their interactions with the physical and psychological world, my hunch is that they lack the moral emotions or moral senses. They lack the capacity for empathy, sympathy, shame, guilt, and loyalty. The reason for this emotional hole in their lives, I believe, is that they lack a fundamental mental tool: self-awareness. Although there is some evidence that animals have self-recognition, recognizing and distinguishing their own bodies from others, there is no evidence that they are actually aware of their own beliefs and desires, and how they differ from those of their peers; future work could, of course, show that my hunch is completely incorrect. Without self-awareness, the kind of empathic response that appears to underlie some of the experimental results described is impossible. Empathy requires not only a sense of self, but a sense of self that ties into what it would be like to be someone else. Empathy represents an emotional fusion of self and other.

If my claim about self-awareness is correct, then animals must also lack a deep understanding of death. To understand death as a system of beliefs, as opposed to simply responding to dead things, individuals must have a sense of self-awareness. Without an understanding of death, they cannot exhibit moral outrage when one individual kills another, though they may well feel a great loss. Feeling a loss and understanding what it means to die are two different things. What, then, do animals know about the process of dying and the categorical distinction between life and death?

Death ties into an understanding of morality on a number of levels. For one thing, understanding death reveals a deep sense of time, a sense of what the past has been like and what the future holds. Further, many of our laws are designed to protect others from being killed or killing themselves, because we see life as precious. To appreciate such laws and support them, one must have an understanding of death.[13]

Understanding what animals know about death can be reduced to the problem of whether animals understand the difference between living and nonliving things. Ants provide an excellent starting point. The sociobiologist Ed O. Wilson noticed that when ants find a dead colony member, they drag it out of the area and then drop it off. The signpost for "death" in the ant is its smell. In particular, when ants die they emit a chemical called oleic acid. It turns out that if you cover a live ant with oleic acid, colony members deposit this individual out of the nest area. Ants seem to equate oleic acid with death, and leave no room for exceptions. Moreover, when an ant dies, there is no indication from the behavior of other colony members that they have experienced a great loss; of course, given that we know so little about the emotions of insects, we are on shaky ground here. If "death" means no more than a drop of oleic acid, then ants lack an understanding of death. They can distinguish between dead and living things, but can be readily fooled by a mischievous experimenter.

To have an understanding of death is to have specific beliefs about what it means to be dead. Children under the age of ten tend to believe that dead is simply the opposite of living, that is, not living. In this sense, inanimate objects are dead, and so are animate objects that have their eyes closed and fail to move. As the developmental psychologist Susan Carey has pointed out, some time after the age of ten, children undergo a conceptual transformation that provides them with a new way of thinking about life and death. This conceptual transformation teaches them that only animate things can die, that upon dying, the brain no longer works, that if they die the same things will happen to them, that killing is bad, and that they have the potential to end another's life as well as their own. In this sense, a psychologically interesting representation of death depends on self-awareness and a richly textured set of beliefs about the meaning of life.[14]

Wilson's work shows that for an ant, death = oleic acid. We might not expect much more psychological sophistication from ants, even though they live in intricate social networks. But what about highly social mammals that develop close bonds and, often, lifelong relationships that depend on cooperation and support? One might imagine that in such societies, the loss of a relative or close friend would be traumatic. If so, there is a rich emotional texture underlying loss, and this could fuel a richer understanding of death.

Field biologists have recorded dozens of observations of elephants and primates responding to dead group mates.[15] Elephants return to a dead group member for days, even months, and handle the bones. Nonhuman primate mothers will often carry their dead offspring around for days, even though the smell of rotting flesh and the lack of responsiveness must indicate the loss of life. Some have claimed that chimpanzees produce a moaning cry in response to a dead community member, a response that many have interpreted as a sign of grief. A response to an object, however, provides only one dimension of our analysis of animal emotion and thought. We need to understand what the animal feels, what it thinks others feel, and what it does to integrate the subjective and objective dimensions of feeling in such contexts.

Consider an observation by Cynthia Moss, the world's leading expert on elephant social behavior. In this powerful description, several elephants are responding to a dying member of their family, a female named Tina:

> Teresia and Trista became frantic and knelt down and tried to lift her up. They worked their tusks under her back and under her head. At one point they succeeded in lifting her into a sitting position but her body flopped back down. Her family tried everything to rouse her, kicking and tusking her, and Tallulah even went off and collected a trunkful of grass and tried to stuff it into her mouth.[16]

Following these events, several individuals returned to the site, covered Tina's body with branches and grass, and repeatedly felt the carcass with their foot or trunk before moving off.

This is a wonderfully rich description. Authors such as Frans de Waal, and especially Jeffrey Masson, have interpreted this observation as evidence of empathy and an understanding of death. Although this is one possible interpretation, there are alternatives. What can we actually infer about an elephant's feelings and thoughts about death from its response to a dead or dying elephant? The first step is to work out the kind of objects and events that elicit the particular response observed. Thus, do elephants respond in the same way to a sick but living member of the family? Would they attempt to prop his or her body up? If so, then the same kinds of behavior have been directed at two different kinds of individuals, one dead and one sick. Do individuals with special social relationships to the dead show different kinds of responses to the bones than unfamiliar individuals? Until we have answers to these questions we can't say much about an elephant's feelings or thoughts about death, although it is certainly possible that they feel grief and a deep sense of loss.

I do not question the observations of animals responding to dead group members. What I question is their interpretation. Given the lack of evidence for self-awareness as well as the capacity to attribute mental states to others, my own hunch is that no animal will be found to have a system of beliefs about death. Animals will care for those who fail to respond like normal living individuals, and may even show signs of depression when a mate or relative has died. However, caring for others, whether they are living, dying, or sick, is only one step in our analysis of the underlying emotion, what animals feel when they see someone who has been injured or is dead.

MOTIVATIONAL BRAKES

Have you ever run into a glass door? I have. It is embarrassing. Typically, one tries to hide such miscalculations as swiftly as possible. Insects and birds also smack into glass doors and windows; unlike humans, however, they often repeat this action without any sense of embarrassment. Insects and birds just don't understand glass or other clear human inventions such as Plexiglas and cellophane. These species did not evolve in an environment with transparent barriers to movement. Neither did we, which is precisely why we are sometimes vulnerable to perceptual errors. But we evolved a psychological mechanism that allows us to correct for such accidents and avoid walking into glass doors over and over again. This mechanism—the power to inhibit actions motivated by strong emotions or drives—develops slowly over the course of human development, can be damaged as a result of injury to a certain region of the prefrontal cortex, and is only weakly developed in nonhuman animals. Most important, the ability to inhibit actions provides the backbone for all moral systems because it is an essential ingredient for problem solving. In its absence, many of our selfish tendencies would take precedence, leading us to act on our passions. Rational thought depends on the ability to evaluate one's options, assign emotional weight to each, and then inhibit or reject one or more alternatives in favor of another.[17]

In the 1980s the developmental psychologist Adele Diamond developed a simple yet elegant test to determine when in their development humans acquire the capacity to inhibit one action in favor of another. Diamond's test involved presenting infants with a transparent Plexiglas box, open on only one side. Inside the box, clearly visible, was a toy the child had never seen before. Novelty is interesting to young infants, and draws both their attention and a reaching response. On some trials, the opening was straight ahead in front of them. When they reached for it, they obtained the toy. On other trials, however, the opening was on the side or on top. This required them to take a detour, reaching around to a side. It also required the ability to inhibit the natural tendency to reach straight ahead. Infants less then nine months old reached straight ahead on all trials, and did so repeatedly without any evidence of learning from their mistakes. Like insects and birds banging into windows, they appear to have in mind a simple yet rigidly fixed rule for reaching: if a desired object is in front of you, reach straight ahead. Even in the face of several failed attempts, they repeat the same action.

Diamond extended this work to rhesus monkeys to evaluate whether the child's inability to inhibit a reaching response is shared with other species, as well as to localize the brain mechanism underlying our capacity to inhibit. Like infants less than nine months old, two-to-four-month-old rhesus monkeys also fail to inhibit the straight-ahead reaching response, and so do adult rhesus monkeys with lesions to a particular region of the prefrontal cortex. Human infants over the age of nine months and normal rhesus monkeys over the age of four months have no problem reaching around to a side, even if they bang into the Plexiglas front panel on the first encounter. These individuals have acquired the capacity to inhibit the natural tendency to reach straight ahead for something that is directly within the line of sight.[18] This capacity depends on a mature and healthy prefrontal cortex, as evidenced by the inadequate performances of infant humans and rhesus monkeys, as well as adult rhesus with this piece of brain removed. Inhibition is therefore crucial when an apparently natural, common, or even rational action is inappropriate and thus must be rejected in favor of an alternative.[19]

If inhibition requires putting the brakes on one option in favor of another, then alternatives must be available. Perhaps the problem for young humans and rhesus is that they lack alternatives? Unlike scientists who have refined the ability to generate and test alternative hypotheses, perhaps human infants and some animals are simply limited to a narrow range of instinctual responses. To explore this possibility, my students Laurie Santos and Brian Ericson ran a slightly modified version of Diamond's box test using adult cotton-top tamarins, and food as opposed to a toy as the target reward. One group was tested with the same procedure and conditions that Diamond used with human infants and rhesus monkeys. Given that our tamarins were adults, with fully mature brains, we expected them to solve this problem by reaching around to the open side. To our surprise, however, the adult tamarins acted like human infants under the age of nine months, two-to-four-month-old rhesus, and adult rhesus monkeys with lesions in the prefrontal cortex. Upon seeing the food enclosed in the box, the tamarins reached straight ahead irrespective of the placement of the opening. This suggests that the capacity for inhibition is affected not only by the developmental history of the individual, but also by the evolutionary history of the species. Adult tamarins, quite simply, appear to lack the kind of inhibitory control that adult rhesus possess, at least in this particular experimental task. Whether this represents a difference between the brains of Old World (rhesus) and New World monkeys (tamarins) is currently unclear, and awaits additional tests with other representative species. These studies nonetheless reveal a significant constraint on tamarin problem solving.

A slightly modified version of the same test was run on a second group of tamarins. In the first session, they were presented with an opaque rather than transparent box. A piece of food was placed on a stage and the box placed on top. To solve this problem, the tamarins had to hold in mind the position of the currently hidden piece of food and then find it by locating the opening in the box. Because the food is no longer directly visible, the motivation to reach straight ahead should be reduced. The tamarins readily solved this problem. On the second session, we presented a transparent box with the food clearly visible. If the first group's problem was due to the visibility of the food—its motivational pull—then the tamarins should nonetheless have problems inhibiting the tendency to reach straight ahead for food that is in view. In contrast, if the opaque box session provided the tamarins with alternative ways of reaching, then they should have little problem with the transparent box. The tamarins performed as well with this box as with the opaque one, which suggests that the sight of food was not the only problem. Having learned about alternative reaching responses with the opaque box, the tamarins applied their knowledge to the transparent box. Moreover, they inhibited a seemingly potent tendency to reach straight ahead for food lying directly in front of them.[20]

The box experiment reveals that animals may have difficulty with a problem when the apparent solution involves a natural, common, or typical response. At the root of such difficulties are powerful motivational drives, emotions that may overwhelm the inhibitory process. However, the box test involves interactions with physical objects rather than objects with minds. To prove that inhibition plays an integral role in moral action, we must show that the same problem surfaces in the context of social interactions.

In chapter 3 we encountered a socially relevant inhibition problem, one that was carefully disguised in the descriptive language of numbers. The psychologist Sally Boysen ran a social exchange experiment with two of the smartest chimpanzees on the planet, Sarah and Sheba. In this experiment, one chimpanzee played the role of selector and the other played the role of receiver. The selector's role was to choose between one of two possible food wells, where one well consistently offered more food than the other. Whichever well the selector pointed to first, the receiver obtained; the remaining food well was left for the selector. Thus, if the selector is selfish—and one certainly imagines that chimpanzees would be under such circumstances—then she should point to the tray that she doesn't want. For example, given a choice between one or four food treats, Sarah should point to the well with one piece, thereby keeping four pieces for herself. Even after several dozen trials, Sarah pointed to the larger quantity, and so did Sheba when the roles were reversed. Neither chimpanzee could inhibit the tendency to reach for the larger food quantity. In this sense, they are like children under the age of four years, as well as many autistic children, who also point to the larger food quantity in similar tests.[21] When Boysen presented Sheba with the same experimental setup, but with Arabic numerals replacing the actual food, she succeeded in pointing to the smaller of the two numbers.

Sheba's performance with Arabic numerals appears analogous to the tamarins' performance on the opaque box test. When she is allowed to use an alternative way of thinking about the problem, she can inhibit the most natural and obvious response of reaching for the larger quantity of food. It is presently unclear whether Sheba's success comes from her ability to understand Arabic numbers or the fact that the cards cover up the food, or both.

It is surprising that two such exceedingly smart chimpanzees could not acquire the simple rule "Point to the food well you don't want." Could their ability to inhibit be so weak? The psychologists Alan Silberberg and Kazuo Fujita have argued that the chimpanzees' failure to pick the smaller quantity is due to a quirk in Boysen's experimental design: regardless of their choice, they are always rewarded. If they pick four pieces of food, they receive one; if they pick one, they receive four. There is no significant cost associated with making one choice or another, especially for two well-fed chimpanzees who presumably receive numerous treats throughout the day. To examine this possibility, Silberberg and Fujita ran a comparable test with Japanese macaques. In the first condition, a human played receiver, offering the subject four treats in one hand and one in the other. Whichever hand was selected, the subject was offered the alternative. This condition was identical to Boysen's test with Sheba and Sara. All three macaques performed the way the chimpanzees' did, consistently selecting the hand holding four treats, and consequently receiving only one treat. In the second condition, the same quantities were presented, but if the subject selected the hand holding four treats, it received nothing at all; if it picked the hand holding one treat, it received four. The macaques solved this problem, consistently picking the hand with one treat. This suggests that the chimpanzees' failure to generate the appropriate solution may have less to do with the problem of inhibition than with an experimental procedure that imposes relatively little cost with respect to the selected response.[22] This explanation does not, however, account for Sheba's success in using Arabic numerals, a task that preserves the original, no-cost design quirk. The fact that Sheba solved this problem shows that she can learn the rule "Pick the quantity you don't want," is motivated to obtain the larger food quantity, and is able to inhibit the natural tendency to reach for the larger food quantity. It is currently unclear whether these abilities are due to Sheba's recently acquired capacity to manipulate numerical symbols or a more general capacity to generate alternative solutions to a problem.

Many animals, including human infants, have difficulty solving problems because they lack the capacity to inhibit powerful response tendencies or

biases. Some problems are difficult to solve because along the path to finding a solution one must suppress motivationally powerful drives that derail
the problem-solving process. The failure of chimpanzees and Japanese
macaques to find the rule "Pick the food reward you don't want" represents
a simple example. I call this form of repetitive error or perseverance "Cartesian," reflecting Descartes' conviction that rationality requires control over
one's passions. When such control fails, Cartesian perseverance can emerge.
Other problems are difficult because the most obvious or rational response
represents an inappropriate solution, whereas an apparently irrational
response is correct. Here, the failure lies in an inability to incorporate new
empirical evidence into a new theoretical perspective of the world.

A common pattern in the history of science is for certain theoretical positions or paradigms to dominate for particular stretches of time. A paradigm's
tenure depends, as the philosopher Thomas Kuhn argued, on the extent to
which it accounts for the data. When the data can no longer be explained by
the theory, and when an alternative theory provides a better account, then a
paradigm shift occurs. The transition between paradigms is, needless to say,
difficult given the fact that most scientists reared in the previously fashion-
able paradigm are resistant to change. Scientists will find it difficult to
inhibit their tendency to interpret the world using the theoretical concepts
with which they are most familiar.[23]

In the first part of the book I argued that both human and nonhuman animals are equipped with three domain-specific mental tools. Each mental
tool operates on the basis of a set of core principles that reflect statistical
regularities of the world—objects cohere as entities over space and time;
physical objects move if and only if contacted; landmarks represent stable
properties of the spatial environment. Like scientific paradigms, these mental tools provide powerful explanations for how things should work. Their
power derives from the fact that they have been road tested over millions of
years by millions of species that have survived and passed down the relevant
genetic material for such adaptive mechanisms. If the principles underlying
these mental tools are like paradigms, then they too should be resistant to
change.[24] If this logic is correct, we are well positioned to generate and test a
simple yet powerful prediction: animals should have the greatest difficulty
solving those problems that require a theoretical shift away from the core

principles. They should have difficulty inhibiting whatever predictions the domain-specific mental tool generates. As a result, they will persevere with the same response, repeating an action over and over again. I refer to this type of error as "Kuhnian," placing it in explicit opposition to Cartesian perseverance, and in honor of the philosopher of science Thomas Kuhn, who developed the idea of paradigm shifts in science.

One of Newton's key insights was to ask why, when released, an apple will fall to the ground rather than rise to the sky. The answer, of course, was gravity, a crucial property of the physical world and a core feature of falling objects. As adults, however, we know that not all objects fall straight to the ground when released. Some objects are thrown off course by other physical objects, while others are filled with helium and thus rise to the skies. In 1995 the developmental psychologist Bruce Hood developed an experimental procedure to test human children's understanding of falling objects, a procedure that my students and I have also applied to cotton-top tamarins. The apparatus consisted of a rectangular frame, standing upright, with a series of three short pipes on top (pipes A, B, and C) and three small boxes (boxes 1, 2, and 3) lined up below; the inside of the frame was empty.

An experimenter presented the apparatus to a subject and then attached an S-shaped opaque tube from pipe A to box 3. While a two- to three-year-old child or adult tamarin watched, the experimenter dropped an object down the tube and then allowed the subject to search for it; children searched for a ball while tamarins searched for a piece of food. If children and tamarins understand the fact that objects travel down the tube—an invisible displacement—then they should search in box 3. Consistently, both the children and the tamarins searched in box 1, the box located directly below the release point at pipe A. This is surprising, of course, because there is no connection between pipe A and box 1. Not only did the children and tamarins repeat this error once, they repeated it twenty to thirty times. This is shocking given the fact that the tube's position remains the same across trials, and the actual position of the ball or piece of food is indicated after each failed search. Even more extraordinary are the following three observations. First, once subjects reliably pick box 3, they don't apply this newly acquired knowledge to new tube configurations, such as a connection between pipe B and box 1. Here, subjects typically search in box 2, directly below pipe B. Second, when the opaque tube is replaced by a transparent one running from pipe A to box 3, subjects now search in box 3; being able to see the ball move through the tube apparently helps. However, when the opaque tube is put back in position, these same individuals return to their old error, searching in box 1; they don't apply the knowledge acquired from the test with the transparent tube condition. Third, although this task appears to require individuals to predict the movement of a hidden object, they could obtain the solution without even watching the ball/food drop. By recognizing the fact that the ball/food always appears in the box associated with a connected tube, one can solve the problem. To eliminate this solution, one must introduce a second tube, a distracter that forces a search for the relevant connections. Children over the age of three years solve both the single and two-tube task, whereas adult tamarins do not.[25]

These results suggest that adult tamarins and children under the age of three years fail this task because of the strong theoretical intuitions or principles that they bring to the problem, principles that generate predictions about the nature of falling objects. Specifically, these individuals appear to expect all falling objects to fall straight down, guided by the forces of gravity. They are making a gravity error. Although we don't understand how much experience with falling objects is required to grasp this intuition, it is clear that it is part of the child's mental toolkit by the age of three, and apparently never becomes part of the tamarin's mental toolkit. Because the intuition about falling objects is so firmly in place, young children and adult tamarins are vulnerable to Kuhnian perseverance, repeating an action because the theory drives them to do so.

The experiments described in this section are relevant to our concern with the evolution of moral systems because they reveal two significant mental limitations among animals and immature humans. Specifically, I claim that young children and all animals have no role to play in shaping the moral community because they have an impoverished capacity for inhibition and conceptual change. To make an ethical decision—to decide between right and wrong—we must choose between two or more alternative possibilities. Sometimes the most tempting possibility is the wrong thing to do. Consequently, we must often inhibit our desires in order to do the right thing. The capacity to inhibit therefore plays a crucial role in moral action, and animals and young children are limited in this capacity. Furthermore, as legal systems change, we must change our conceptual framework—we must engage in paradigm shifts. The famous case of *Roe v. Wade* rocked the moral fabric of many U.S. citizens and forced them to change their views about women's rights. Again, young children and all animals are limited in their ability to engage in conceptual change and thus cannot understand why abstract rules and regulations are created, and sometimes change.

FAIR IS FAIR

The problem of altruism stumped Charles Darwin. His theory could not explain why some individuals, in some species, behaved in ways that were costly to their survival and reproduction, yet beneficial to those who received their generous gifts. Why, for example, would an adult scrub jay help its parents rear the current brood of nestlings and thereby forfeit the opportunity to reproduce? Why do social insects such as bees and wasps consist of sterile individuals, leaving the work of reproduction to the elite queens? The answer to this problem was long in coming. When it arrived, almost a hundred years after Darwin considered it, it revolutionized biology. Not long thereafter, these ideas revolutionized—or at least challenged—psychology, sociology, economics, politics, and even the arts.[26]

The answer, developed from about 1965 to 1975, was delivered by the evolutionary biologists Richard Dawkins, William Hamilton, John Maynard Smith, Robert Trivers, and George William.[27] The crucial insight was to see altruism as selfish and examine the evolutionary function of an interaction in terms of genes, the fundamental replicators of life. If selection favors gene replication, then there are several ways to win. Reproducing on one's own is the most direct approach. A more indirect way is to help out those who are genetically related. Because I share half of my genes in common with my mother, I win in the game of gene replication if I help her raise future siblings, who share half their genes in common with me; the degree of genetic overlap with siblings depends, of course, on whether we share the same father. Using this logic, the seemingly costly actions of scrub jays and insects can be explained. Scrub jays that fail to reproduce in their first year are nonetheless contributing to the gene pool by staying at home and helping their parents feed and defend the current brood. If these *helpers at the nest*, as they are called, help their siblings survive and reproduce, then they have passed on a package of genes to the next generation. In the case of insects, sterile workers that defend the colony, build nests, and tend the young gain by contributing indirectly to their relative's reproductive output.

Why, however, would anyone ever help a genetically unrelated individual? The standard answer before the idea of selfish genes came to fruition was that this kind of unselfish altruism occurred because it benefited the group or species. Alarm calls were given to protect the group. Food calls were given to attract group members and thereby prevent their starvation. Pure kindness without any hope of recompense. The more modern, and generally accepted, view is that altruistic actions toward unrelated individuals are motivated by an understanding of reciprocation, an "I'll groom your back if you'll groom mine" attitude.[28]

To understand the problem of reciprocal altruism, consider the following hypothetical scenario, a snapshot of early hominid life. Fred and Barney are out on the savanna looking for dinner when they run into a sabertooth tiger. Both Fred and Barney know that they can kill the tiger if they fight together; neither can survive on his own. They must cooperate. There is, however, a cost associated with fighting. Although the sabertooth tiger will either run away or die, Fred and Barney could get hurt in the process. Consequently, both men will be tempted to escape before the battle ensues. This tension between cooperation and defection has been called the prisoner's dilemma by economists interested in the dynamics of games, and has been explored in great detail by economists, computer scientists, and evolutionary biologists.[29] The outcome of this work is an understanding that if reciprocal altruism is to evolve, several conditions must be met.

Trivers originally flagged three crucial preconditions for reciprocal altruism to evolve. First, there must be a cost to giving and a benefit to receiving. If the donor is flush with resources, giving is cheap and therefore not altruistic. If the recipient can do without the handout, there is no benefit. Fred and Barney's encounter with the sabertooth certainly satisfies this condition. Second, there must be a delay between the first bout of altruism and the reciprocated act. After the donor gives, the recipient must return the favor within a reasonable amount of time. If Fred helps Barney kill the sabertooth, there will certainly be occasions in the future for Barney to return the favor. Third, giving must be contingent on receiving. Fred is going to help Barney only if he thinks that Barney will return the favor.

What is interesting about these seemingly straightforward conditions is that they lead to several other conditions, each requiring more sophisticated psychological capacities from the donors and recipients. It is precisely because of these requirements that we have only a handful of cases of reciprocal altruism among animals.

Imagine that Fred and Barney are encountering the sabertooth for the first time. They both know, individually at least, that to avoid being eaten, they must either both run away or both fight. If they run, however, the slower of the two may get eaten. Thus, there is an incentive for both to cooperate or for both to dash away. Cooperation is favored if the odds of encountering this situation in the future are high. In contrast, the motivation to flee is favored if the odds of encountering a sabertooth are low, if the odds of the two meeting again are low, or if either Fred or Barney have a bad reputation, one that implies that he is unlikely to repay his debt. Reciprocal altruism therefore requires a substantial insurance policy. There must be some guarantee that there will be opportunities to interact in the future. Given that there will be such opportunities, individuals must be able to evaluate the benefits of the resources exchanged, the costs associated with giving, and the reputation of the potential partner. Such calculations, which can happen without the animal being aware of them, are critical because reciprocal altruism is based on a system of exchange, one that depends on some agreed-upon currency. For example, if Fred helps Barney out against the sabertooth, Barney must reciprocate in the future when Fred is in trouble against a sabertooth. But would Fred also accept help in another context or in another currency? For example, would Fred accept food from Barney during times of scarcity and would this count as a fair exchange?

Looking carefully at the conditions above, we can derive the kind of mind capable of reciprocal altruism. Individuals must be able to recall prior encounters and set up a timer to determine by when an altruistic act should be reciprocated. Individuals must be able to recognize each other and recall what they gave and what they received. They must be able to place a value on every altruistic action, a value that links the costs of giving with some effect on survival and reproduction. Given the economics of exchange, individuals should be able to detect and punish cheaters, individuals who reneged on an exchange. If reciprocal altruism is selected for, and cheating selected against, then the former should be associated with feeling good and the latter with feeling bad; cheating should also be associated with feeling afraid since cheaters will, if caught, be punished.

That is a lot of psychological gear. Although certain types of reciprocal exchange may demand fewer or more psychological capacities, the set described above sets the boundaries we need to explore in looking for cases of reciprocal altruism in the animal kingdom.

In the more than twenty-five years since Trivers's theoretical analysis of reciprocal altruism, there have been numerous attempts to provide empirical support for these skeletal ideas. For example, to beat up a more dominant opponent over access to a female, two lower-ranking male baboons will join forces. On any given cooperative venture, only one of the males will win the sexual prize. However, given the opportunity for repeated interactions, coalition partners appear to take turns in their attempts at sexual exploitation. Dolphins do the same, but take the complexity of the situation to dizzying heights. Like the United Nations armed forces, characterized by coalitions within countries as well as those between countries, dolphins also form coalitions within coalitions. Thus, one group of male dolphins will help another group of male dolphins herd a female away from a third group of male dolphins. We don't yet know whether these super-coalitions reciprocate with respect to gaining access to females, but the intuition is that they are trading favors. In bonobos, females exchange food for sex: if a male has access to food that a female wants, the female will approach, offer sex, and then take the food away while the male mounts and copulates. Unlike the arrangements of dolphins and baboons, this trade involves different currencies. Scientists have yet to work out whether a fair exchange involves one copulatory bout for one bundle of food, or whether the exchange rate can vary, depending perhaps on the current market value of resources.[30]

The examples above suggest that cooperative interactions require sophisticated computational abilities, keeping track of who is dominant to whom, who will be supportive in a fight, and who will turn the other cheek. In general, one finds that among species forming coalitions and alliances, a large brain is part of the toolkit.[31] We don't see coalitions in insects, amphibians, or birds, whereas we do see them in many of the carnivorous cats and dogs, as well as in dolphins and primates. Although we know that brain volume is not the only metric of how smart a species is, and may in some cases say little about computational power, one might expect such brainy species to provide compelling examples of reciprocal altruism in action. Surprisingly, perhaps, the award goes to vampire bats and guppies, species that tend to miss out on the trophies for cleverness. Since studies of both species have provided comparable data with respect to our concerns in this chapter, I focus below on the bats.[32]

Vampire bats live in large social groups, recognize each other by voice, and live long enough to develop stable social relationships. The recipe for survival is simple: go out each night and find a blood meal. The problem is that blood is often difficult to find. Although bats can go without a meal for a couple of nights, they will starve to death after sixty hours. A bat with a stomach full of blood, however, can regurgitate some to another in need. Regurgitation therefore represents an insurance policy against starvation. These details place vampire bats in a prisoner's dilemma, resembling the situation faced by Fred and Barney. Blood is the currency. It is a limited resource, of value to all. Giving it up is costly. Receiving it is beneficial, especially as the sixty-hour starvation mark approaches. Do bats cooperate and help each other out? Do they punish those who fail to reciprocate?

To determine how vampire bats respond to the prisoner's dilemma, the evolutionary biologist Gerry Wilkinson analyzed their pattern of blood exchange. Individuals were most likely to regurgitate to those who had regurgitated in the past. Crucially, this pattern of exchange occurred among individuals that were either genetically unrelated or distantly related. A selfish gene perspective cannot explain this pattern. Regurgitation does not flow along kin lines, so individuals cannot be incurring the cost of a blood donation to help promote their genes. Rather, as Trivers's theoretical arguments predicted, individuals will act altruistically toward nonrelatives when they expect giving to be reciprocated in the future, or when it has been reciprocated in the past. Wilkinson further observed that bats punished cheaters, rejecting requests for blood if those asking had failed to donate blood in the past. How bats detect cheaters is still somewhat of a mystery, though Wilkinson suggests that an individual's capacity to regurgitate can be estimated from its stomach size. When a bat obtains a blood meal, its stomach distends. By grooming each other, individuals can see and presumably feel the stomach, and thus estimate the potential for regurgitation.

Vampire bats fulfill all of Trivers's conditions for reciprocal altruism: bats recognize each other, are long-lived, have repeated interactions with others and count on them in the future, depend on the exchange of a resource for survival, and have the capacity to detect cheaters that attempt an end run around the system of exchange. In terms of their thoughts and emotions, however, we are left with several puzzles:

Do donors expect recipients to return the same volume of blood, some blood, or an amount that represents a compromise between what the donor wants and what the recipient can give immediately?

Do bats calculate the precise amount of blood given and received, or simply generate a rough estimate?

Do donors establish a rigid or flexible time frame for returning the favor?

Does punishment work, causing cheaters to adhere to the payback agreement that has evolved as a convention?

Are cheaters aware of the potential consequences of their violation, aware that they may be denied a regurgitated blood meal when they really need it?

Does cheating represent a failure to return the exact amount of blood received in the past, a failure to return some blood in a timely fashion, or some combination of these factors?

Do cheaters develop reputations that are recognized by both individuals that have been cheated and individuals that have merely observed their failure to uphold the deal?

Do bats feel sympathy or empathy toward individuals on the verge of starvation and consequently regurgitate even if they have never received blood from them in the past?

Are older and younger individuals who may have greater difficulty finding a meal treated compassionately, and provided with more blood than they are expected to return?

We don't have answers to these questions. An evolutionary psychology of reciprocation must, however, address these problems, especially if work on animals hopes to shed light on reciprocation and cooperation in humans.[33] For some problems it will be difficult, perhaps even impossible to provide answers. We don't, for instance, have any way of directly measuring the amount of blood consumed or regurgitated by a bat. Consequently, the *fairness* of an exchange is difficult to evaluate. We can only assess the consequences of regurgitation, seeing whether individuals are denied a blood meal or given only a small one (perhaps measured in time spent regurgitating) when they failed to regurgitate in the past. Other questions can, however, be answered. We can measure the time elapsed between reciprocal exchanges, how often particular individuals are given preferential treatment or special status, and the causes of punishment. We can also look to other species for answers.

Lion groups consist of genetically related adult females, their offspring, and a few adult males. Females respond aggressively to threats from other groups, and the intensity of their response appears to depend on the relative power or strength of the opponent (see chapter 3). Karen McComb's playback experiments show that females respond aggressively to the roar of a single foreign female, but retreat when they hear roars from three females. To explore the possibility that female lions cooperate during group defense, the biologists Robert Heinsohn and Craig Packer conducted playback experiments with lion prides living in Tanzania. Upon hearing roars from unfamiliar lions, females responded in one of four ways. Some females were consistently leaders, marching off toward the speaker on each trial, independently of what other pride members were doing. Diametrically opposed to these lionesses were the individuals who consistently lagged behind. Between the extremes were the females whose behavior depended on what others were doing. Some lagged least when they were needed most. Others lagged most when they were needed most. Given the fact that leaders were more likely to be attacked than laggards, Darwinian theory suggests that there should be strong selection against cheaters: their pride mates should catch and punish them for their laissez-faire attitude. Although unconditional leaders were skeptical of laggards, looking back at them following a simulated territorial intrusion, they nonetheless walked off on their own toward the speaker. Thus, leaders either arrived alone at the speaker or arrived well in advance of any laggards.

Heinsohn and Packer's experiments show that lions fail to cooperate under these particular experimental conditions. There is no system in place to punish laggards, and thus, lion prides function with a mixture of leaders and laggards. As in our discussion of vampire bats, we are left with several questions concerning lion psychology, what lions think when they confront danger:

> What makes an individual a leader or a laggard? Genetic analyses suggest that the personality of leader or laggard is not inherited. What aspects of the environment therefore trigger such personalities and allow for their stability over time?

Why are laggards tolerated rather than punished for their laziness?

Are laggards more likely to lag if the direct costs to them are low (e.g., they have no cubs, and thus are not vulnerable to attack from infanticidal males)?

Given that a lion pride requires at least some leaders in order to defend itself, what determines the relative proportions of leaders and laggards in a group?

The lion data, though incomplete, suggest that animals have reputations and that some individuals take on certain roles, based presumably on special skills. Of course we don't know whether laggards *value* or *respect* leaders, whether they see them as special in some way. Nonetheless, this situation reveals an asymmetry in the flow of helping, and raises the question whether laggards have skills that leaders lack and whether laggards help leaders in contexts other than those of intergroup threat. Though we don't have answers to such questions for lions, experiments on monkeys and apes provide preliminary insights into the general issue of whether animals value those with special skills.

While I was conducting my thesis work in Kenya during the 1980s, one of my vervet monkey groups suffered a bout of intense predation and all the adult males died. This situation represented an exceptional opportunity for an adult male—first dibs on a group of receptive females. Although several male suitors passed through, one male—Macaulay—stayed. Macaulay had lived his entire life in an adjacent group and thus had ample time to watch his neighbors. He also had time to assess the quality of resources in his own group as opposed to his neighbor's. Leaving his home meant gaining exclusive access to females but losing out on a resource-rich territory. The neighbors lacked a source of permanent water, and had access to a limited range of high-quality foods. After several months in his new home, Macaulay went on several solo expeditions back to his old territory. Upon arrival, he stayed for supper and a drink. He then headed back, walking several hundred meters across open and extremely dangerous grasslands. One day, he made a foray back to his old territory, looked around, walked back without eating, and then proceeded to lead the rest of his group into his old, rich habitat. Macaulay led his troop into this area for months, and often encountered his old group. Rather than being friendly, they attacked him. But Macaulay attacked back, supported by the females as well.

When I returned from the field, I mentioned this story to a colleague who worked on fly genetics, emphasizing the point that individuals are extremely important in primate societies. My colleague laughed, asking me how I could possibly show that Macaulay was personally responsible for these events. I then provided him with the sad ending to this story. One day, Macaulay left his group on a trek back to his old territory and never returned. We assume he was eaten by one of the many predators that prey on vervets. Although Macaulay's group clearly knew the route to nirvana, they never returned on their own. Other males entered and took over, but they never made the trek. Macaulay appeared to have a special character as a fighter and a leader.

In 1988 the ethologist Eduard Stammbach set up an experiment with long-tailed macaques to test the idea that individuals with special skills should receive special treatment from those lacking in such skills. All individuals in a social group were trained to press a lever on a machine whenever a light was illuminated. Subjects were rewarded for their efforts with a popcorn treat. Once everyone knew what to do and when, subgroups were created, and the lowest-ranking member in each subgroup was trained to press a set of levers in a specific sequence. Doing so caused the machine to deliver enough popcorn for three individuals.

At first, the high-ranking individuals threatened the low-ranking individuals away from the dispenser, thereby intimidating them. The high-ranking individuals learned, however, that the low-ranking individuals had a unique skill and followed them to the machine, waiting to grab all the popcorn. As a result, low-ranking specialists stopped working the machine. But this strike didn't last long. Some higher-ranking individuals changed their interactions with the specialists. Rather than chase specialists away from the dispenser, they inhibited all aggression, approached peacefully, and allowed them to eat a portion of the popcorn delivered. Further, high-ranking individuals started grooming specialists more often, even during periods when the machine was inoperative. Although this attitude change enabled low-ranking specialists to access food that would normally be unobtainable, it had no impact on their dominance rank in the group. Specialists kept their rank, but were allowed a moment at high table when their skills were of use to the royalty.

A striking feature of this experiment is the consistency with which non-specialists responded to specialists. The specialist was tolerated at the popcorn dispenser independently of age, rank, or sex. Further, nonspecialists appear to have realized that simply hanging back and waiting for the specialist to operate the dispenser was insufficient. They had to invest in specialists, groom them, and suppress their natural tendency to chase them away from a valued resource. As discussed in the previous section, they had to inhibit their desire to attack low-ranking specialists over access to food; they also had to provide an incentive for the specialist. Clearly, high-ranking animals can inhibit their emotions in this context, as evidenced by their decrease in aggression toward specialists and their toleration of specialists at the popcorn dispenser.[34]

Is it reasonable to conclude that high-ranking macaques learn to respect specialists, placing value on their unique skills? Do nonspecialists tolerate specialists at the popcorn dispenser because they think it is fair? Or is toleration the result of self-interest, the realization that the only route to popcorn is to allow specialists to share in the food that they have worked to obtain? At this point we don't have enough information to test between these two alternative readings. To show that macaques or any other organisms have a sense of fairness or respect, we would have to show that they have a sense of what is right and wrong, and that they act on the basis of such principles. At some level, then, we must be able to show that they adhere to particular rules of order, engaging in behaviors that maintain the peace and prevent chaos, war, and anarchy. They must not only act according to a set of normative rules, but also accompany such actions with a set of normative feelings. Do they?

In 1991 the ethologists Hans Kummer and Marina Cords conducted an experiment with long-tailed macaques to determine whether they have an understanding of ownership, and in particular, a respect for property rights. Though such ideas as respect, property rights, and ownership would seem to be strictly human concerns, in a sense all animals are vested in the issue of property, ownership, and respect. Any territorial animal fits this bill. We know, for instance, that in almost all territorial species, intruders respect territory ownership. If an individual moves into a new area and encounters an individual who defends the area, chances are that this is the territory owner. The intruder therefore retreats to avoid a struggle. The space that a territory owner defends is functionally equivalent to his property, and an intruder's respect reveals his acknowledgment of ownership and property rights. Territorial fights tend to occur in cases where the challenger is significantly bigger and stronger than the resident. For example, the shell on a hermit crab's back represents its property. However, because hermit crabs continue to grow throughout life, they are constantly in the market for a new shell. Thus, when one crab encounters another, they tap on each other's shell to determine the quality of the fit. Tapping returns a sound that indicates the resonance frequencies of the shell. When the owner is small relative to the size of its shell, tapping resonates differently than when the owner is large. If the resonances reveal a poor fit, the challenger can use this information to boot the owner out and thus initiate a shell swap. In contrast, when opponents are well matched, there are rarely successful challenges to ownership. In this sense, hermit crabs respect owners and stay off their backs.

Kummer and Cords tested macaques in a situation where the property— a transparent tube filled with raisins—was either fixed to a wall or free, and was attached either to a long piece of rope, a short piece, or no rope at all. Thus, when the tube was fixed, an animal could monopolize the resource by staying next to it. When it was free, the animal could monopolize the resource by carrying it, but would be more vulnerable to theft when a string dangled behind. Of interest were the cues used by more dominant individuals to assess ownership by a subordinate who was allowed to interact first with the raisin tube. Though dominants can take away resources from subordinates, are there conditions under which they respect the subordinates' possessions? Consistently, dominants took over fixed tubes more often than free tubes, and took over free tubes when the subordinate failed to carry them; staying close to the tube and looking at it were insufficient cues of ownership, at least from the dominant's perspective. Dominants were also more likely to challenge subordinates when the rope attached to the tube was long than when it was short. In the words of Kummer and Cords, dominants appeared "inhibited" from challenging subordinates who carried an object close to them than those who carried it at a distance. Here, then, is a beautiful example of how inhibition plays a crucial role in maintaining conventions, in this case, the convention of ownership or possession.[35]

The establishment of ownership in long-tailed macaques is anything but a simple affair. Dominants do not view subordinates as mere victims, individuals that they can rob of their belongings whenever they so desire. Rather, there appear to be conventions for when dominants can exert their power. Under certain conditions, subordinates are allowed to keep their possessions. Thus, long-tailed macaques have evolved rules for protecting subordinates. Such rules come awfully close to the kinds of property rights that human societies have evolved. The difference, I believe, is that such animals lack an understanding of what constitutes normative behavior (doing the right thing) and what constitutes a violation (doing the wrong thing). As I will argue in the next section, I do not believe that animals have an understanding of right and wrong, though they will certainly defend themselves when their possessions are threatened.

An interesting question to emerge from Kummer and Cords's work is whether an animal's respect for, or failure to challenge, an individual in possession of food applies to the possession of other individuals.[36] Human parents tend to think of their children in a possessive fashion, and certainly the law tends to see young children as property. Similarly, couples often have strong intuitions about ownership, tending to see each other as a kind of property, as something to possess and guard. Such feelings can lead to extremes, including uncontrollable jealousy and obsession about possible infidelities. The metaphors "henpecked" and "on a short leash" are rooted in property rights. Do animals see their pair bonds in a comparable manner? Do bachelors respect other pair bonds or do they perceive them as targets of opportunity, fair game in the battle for gene replication?

One of the primary messages to emerge from recent work on animal mating systems is that true monogamy is a myth. Although many birds and some primates appear to form monogamous pair bonds, each member of the pair moonlights and finds others to have sex with. Even though such sneaky matings are rarely seen by human observers, genetic analyses tell us that they are going on all the time. Females are therefore mating with more than one male, although they provide the illusion of a lovey-dovey pair bond.

Sexual infidelity aside, some animals do appear to respect sexual ownership. The best example comes from studies by Kummer and his colleagues of wild and captive hamadryas baboons. This species lives in a complex society consisting, at one level, of harems—one adult male, several adult females, and their offspring. At a higher level, hamadryas form a large community or population, consisting of dozens of harems. Harem males are rarely challenged over access to their females. Rather, bachelor males attempt to build harems by recruiting juvenile females.

To better understand why bachelors respect harem leaders, Kummer and his colleagues released a male, Sam, into an enclosure with a female, Betty, while another male, Joe, watched in an adjacent cage. Sam immediately approached Betty, followed her around, groomed her, and kept her close by, like a dog herding his sheep. When Kummer released Joe, Joe kept his distance, leaving Sam unchallenged. This kind of situation presumably establishes a dominance relationship between Joe and Sam, with Sam assuming the position of dominant male. To test this idea, Kummer allows Joe to enter an enclosure with a new female, Sue, while Sam watches. Joe makes the same moves as Sam, herding Sue and keeping her within arm's reach. When Sam enters the enclosure, he keeps his distance. Both males appear to respect each other. They seem to have a convention about ownership, one that is based on the premise "first come, first served." Like long-tailed macaques, therefore, hamadryas respect property, but extend the target domain from food to members of the opposite sex.

Rules, conventions, and respect are all fine and good, but sooner or later, someone will try to outwit the system. Such lawbreakers must be punished, and after their punishment there must be a mechanism for reestablishing the peace. At present, we have little evidence that animals punish each other.[37] In chapter 7, for example, we learned that rhesus monkeys were more likely to be attacked by other group members if they failed to announce their food discoveries than if they announced them by calling. On a functional level, one aimed at the reproductive and survivorship consequences of behavior, this aggressive act looks like punishment. Before we conclude that it is punishment, however, we need evidence that such attacks reduce the likelihood of a second or third offense, that cheaters know that they are breaking the rules, and that punishers know that they are attempting to instill societal reform. The same can be said of the vampire bat work. Although bats often fail to regurgitate for those who failed to provide blood in the past, we don't know whether such rejections—a form of punishment—deter future selfishness. One might well imagine that the costs associated with being attacked (or having blood withheld) would function as significant deterrents. At present, however, we simply don't have the relevant data to say one way or the other. And without such data, we are on thin ice with respect to the claim that animals punish one another when a rule is violated.

Even if the capacity to punish has not evolved in animals, there must be a mechanism to cope with the consequences of aggressive interactions.

Highly social species must coexist in the face of fights and squabbles. Such coexistence requires a system to reduce tensions and the day-to-day stresses of living with an enemy or rival. In the 1980s the ethologist Frans de Waal brought this problem into focus with his work on captive chimpanzees. Following up on his analysis of chimpanzee politics—a careful description of the complex strategizing and status striving that chimps exhibit—de Waal focused on the peacemaking capacities of this species, as well as that of primates with different social organizations and temperaments. What he found was a peaceable kingdom, animals that would make up after a fight by kissing, hugging, or holding hands or testicles. In the species observed, making up or reconciling served the function of reducing tensions and allowing animals to get back to the business of life, seeking food, watching for predators, and finding mates.

De Waal's initial exploration into peacemaking among primates set off a veritable cottage industry of work on reconciliation in primates.[38] We know, for example, that almost all primates do it at some level, that when they reconcile they tend to do so immediately, that some species do it more than others, and that highly valued partners will reconcile more than pairs that are indifferent about the relationship. We also know that although stump-tailed macaques reconcile more often than rhesus monkeys, young rhesus can be induced to reconcile at higher levels if they grow up in the company of stumptails. Thus, even though each primate species tends to enter the world with a particular predisposition for reconciling, it is not fixed. Some kinds of experience can turn a tyrant into a loving, caring, and peaceful soul.

What is lacking in this literature is a clear understanding of what individuals feel and think before, during, and after reconciliation. For example, although individuals will often try to resolve their differences after a fight by grooming each other more, holding hands, and feeding together, do they subsequently feel better about the relationship, do they feel calm, do they anticipate future problems, or have they learned how to avoid problems in the future? Some studies show that when two previous enemies act friendly toward each other, each tends to show behavioral signs of relaxation, implying that tension has been reduced. However, one could just as easily turn the correlation around and argue that because tension has been reduced following a fight, this allows the two enemies to make up. Tests have shown that under certain conditions, friendly reunions actually help reduce tension between pairs, allowing them to reestablish a low-stress relationship. These examples, plus the many others that have been accumulated over the past twenty years, allow us to conclude that many primate societies have evolved systems to reduce the day-to-day tensions of living a life in the company of others. This system is certainly an important part of getting along in a society with rules.

AGENTS, PATIENTS, AND RULES

In the opening moments of the computer-animated movie *AntZ*, we find an ant, sporting Woody Allen's voice, lying on a couch made out of leaves. The ant, named Z, is telling his therapist about his angst, the woes of being a worker in ant society, the drudgery of day-to-day life, the constraints on free will, and the oppression of the workforce. Z is clearly worried, knowing full well that what he has expressed to his therapist violates the kinds of normative thoughts and feelings that ants are *supposed* to have. This moment of inspiration, however, ignites a cataclysmic set of events that ultimately transforms the community. He not only challenges the norms of rational action—what one is expected to do as a worker as opposed to a soldier, queen, or other member of the royalty—but also violates the normative feelings that have been genetically assigned to his class. The sophisticated viewer will, of course, have picked up on the fact that about 99 percent of all worker ants are female, rather than male, but so goes the magic of film. Z falls head over heels for Princess Bala (a.k.a. Sharon Stone) the female in line for the role of queen, the one who will inherit the throne and make babies to insure the survival of her genes. The violation of normative feelings and actions is reciprocated as Princess Bala mingles with the working class in the ambiance of a local pub, dances with Z, and ultimately falls in love with him. Z manages to switch roles with his soldier friend Weaver (a.k.a. Sylvester Stallone), another violation of the ant code of conduct. Z leaves the colony with Princess Bala in tow, and then sets the wheels in motion for a revolution against the party line. For a while, the workers and soldiers are deliriously happy as they contemplate the virtues of free will, individuality, and the freedom to express one's wishes and desires. Though Z and Princess Bala prevail, life generally settles back to the ant norm, minus one villainous General Mandible, the mastermind behind the attempted coup.

AntZ rings true to a human audience because we understand the value of free will. We understand the importance of intellectual and emotional freedom, and scorn societies in which such freedom is taken away. Is there any evidence, however, that other animals have free will? Do they respect different styles of thinking and acting? Do they feel as though they have the right to act on their own desires? Do they judge other members of society on a scale that runs from right to wrong? Do they judge their own emotions and actions on a comparable scale and use this evaluative mechanism to motivate future interactions? If the answer to these questions is yes, then in the lingo of philosophy, animals are *moral agents*. If the answer is no then animals are *moral patients*. As moral patients, they deserve to be treated fairly by all moral agents—individuals who have the ability to make ethical distinctions and carry the burden of moral responsibility and duty that societal norms legislate.[39]

The arguments I have been developing in this book suggest that animals are moral patients but not moral agents. To clarify what is at stake, however, we must address six interrelated problems. I list them below in the form of an annotated wish list, a set of desiderata for future research.

First, if animals are moral agents, they must place values on the moral emotions. Animals experience a diversity of emotions—fear, anger, surprise. What is controversial, however, is the subjective experience of emotion, what it is like to feel afraid, angry, or surprised. Some authors think that animals experience the moral emotions, feelings such as guilt, shame, and embarrassment. I don't. Nor do I think animals experience empathy. The reason underlying my claim is that I don't believe animals have self-awareness, a sense of self that relies on a richly textured set of beliefs and desires. To experience the moral emotions one must have self-awareness. The mirror test discussed in chapter 5 shows that animals have some sense of their bodies, but says nothing about their thoughts and emotions. Further, the experiments reviewed in chapter 7 suggest that animals lack the capacity to attribute mental states to others. Together, these data argue against the possibility of self-awareness in animals. Even if I am wrong, and we find good evidence for such emotions in animals, we must solve another problem. Among human cultures, the moral emotions are universal. To be human is to have the moral emotions. However, we not only have them, we also place values on them. Guilt and shame are associated with doing something wrong, violating a norm. Showing compassion toward someone who has been injured or deprived of their rights is viewed as a positive action, doing the right thing. Research on animals must therefore establish that animals have moral emotions and view these emotions in a context of right and wrong.

Second, to act as moral agents, animals must have powerful inhibitory mechanisms that allow them to control their passions and alter their expectations. Under certain circumstances, animals can override their passions, inhibiting powerful desires to mate with a receptive female or feed on a bounty of food. In other situations, animals can inhibit a potential solution to a problem, even when it appears to be the most likely candidate. These two inhibitory mechanisms, however, are limited. As several experiments reveal, animals often fail to act rationally when they are challenged by a motivational lure. This failure to inhibit leads to what I have called Cartesian perseverance, a repetitive response that emerges when the motivational system takes charge. In these situations, the animal's desires overwhelm its capacity to solve problems. It either fails to see the alternative solutions or fails to weigh them appropriately. Thus, chimpanzees fail to override their desire for larger quantities of food, even when they repeatedly fail to obtain the larger food bounty by reaching for it. Sometimes their failure to inhibit a response is due to a failure to generate alternative solutions or hypotheses for a problem. This kind of inhibition problem leads to what I have called Kuhnian perseverance. Here, the animal may well be under control emotionally, and yet it experiences difficulty with a problem because of the strong theoretical intuitions it brings to the task. The tamarin's inability to solve the tube problem provides an example of Kuhnian perseverance. It represents a failure to override the power of a gravity bias. What I suggest, therefore, is that animals have been selected to have innately specified expectations about the physics of the world, as well as the general psychology of members of their social group. Innately specified expectations evolve as a result of statistical regularities, events that are commonly experienced during a lifetime and during the evolutionary history of the species. Having been road-tested, such expectations work well on average, and thus have been selected for. Unfortunately, this kind of rigidity won't work with the general goals of a moral society. Whether such goals aim to regulate behavior, facilitate the greatest good for the greatest number, or something else, an individual incapable of conceptual change is doomed.

Third, if animals are moral agents, they must consider the beliefs, desires, and needs of others when planning an action. At present, we have no convincing evidence that animals attribute beliefs and desires to others. In fact, all the experiments reviewed in chapter 7 suggest that although monkeys and apes are sensitive to how others behave, they don't base their decisions on what others believe, what they desire, or what they can see; all of these conclusions may, of course, change as additional experiments are conducted. Similarly, we also lack evidence that animals have access to their own beliefs, reflect on them, and contemplate how particular events in the future might change what they believe. If this lack of evidence correctly reveals a lack of capacity, then animals can certainly cooperate, beat each other to a pulp, and make up after the war. But they can't evaluate whether an act of reciprocation is *fair*, whether killing someone is *wrong*, and whether an act of kindness should be rewarded because it was the *right* thing to do.

Fourth, if animals are moral agents, they must understand how their actions will impact on the feelings and thoughts of other individuals of their species, and take these consequences into consideration when planning an action. If animals lack the capacity to attribute mental states to others, then they are creatures of the present. This is not to say that they lack the capacity to anticipate or think about the future. They certainly do. What they lack is the capacity to think about how their own actions might change the beliefs of others or cause them pain or happiness, and how risky behaviors such as an attack on a new mother or a competitor might result in their own death, the termination of life and all that it brings.

Fifth, moral agents understand the notions of duty and responsibility and use these principles as guiding lights when interacting with others. A variety of animals help each other by cooperating. Male dolphins join together to form coalitions so that they may gain access to females. Chimpanzees often appear to play different roles in their cooperative attempts to hunt prey. Although I tend to think of such cooperation as selfishly motivated, we don't understand what animal cooperators think about their relationship. Do they feel as though it is their duty and responsibility to help those that have helped them in the past? Do they feel as though they have been treated disrespectfully when their partner has reneged on an offer to help? Having a sense of duty and responsibility is at the core of being a moral agent, and if animals have no sense of self, I don't believe they should be included as members of this class.

Sixth, if animals are moral agents, they must understand the norms of action and emotion in their society and have the capacity to engage in a revolution when their rights have been violated. We know that in many animal societies, dominant animals attack subordinates if the subordinate has access to a resource that the dominant wants. However, if a subordinate walks in front of a dominant and takes first dibs on an estrous female, is the dominant incensed? If the dominant attacks the subordinate, is this because he thinks that there is a code of conduct, one that dictates who has mating rights, first dibs? Would a subordinate ever think about changing the system, overthrowing the normative responses and feelings that define life in a primate group? Animals certainly respond to violations, to individuals attempting to sneak a mating or a piece of food. However, such responses do not seem to be guided by a sense of what is right, a sense that the violation represents an injustice to the group or the species. Rather, when a dominant attacks a sneaky subordinate, it is because the dominant is selfish, and will do what he can to safeguard the resources. Although subordinates may think that this is unfair, I know of no instance where an animal has attempted an overthrow of societal norms. No subordinate has ever built up a coalition of support to derail the system, even though subordinates do overthrow dominants. Following the takeover, the rules are the same, even if the new ruler is kinder or more aggressive than the previous ruler. If there is no understanding of normative responses and emotions, and what constitutes a violation, there is no moral system, and there are no moral agents.

Our planet is inhabited by thinking animals, each confronting the problems of making a living with their own unique mental toolkits.

EPILOGUE

What It's Like to Be a Spider Monkey

In the early 1940s the biologists Don Griffin and Robert Galambos made the astounding discovery that bats use biosonar to avoid obstacles and hunt prey in the dark. After a scientific presentation of their findings, a prominent physiologist approached Galambos, shook him by the shoulders, and told him that no one would ever believe that such a small animal had evolved a sonar system. After all, how could mother nature preempt the navy, which had been working on the problem of sonar for years with nowhere near as much success as the bat? Ultimately, Griffin and Galambos succeeded in gaining the acceptance of their colleagues, with the result that bat echolocation is now firmly ensconced in biology textbooks as a classic example of an evolved specialization.

In the years following the discovery of bat echolocation, Griffin gave lectures to students and faculty at Rockefeller University describing how bats perceive the world. To dramatize his points, he sometimes released one or more bats into the dining hall. The bat's agility instilled both awe and a bit of fear among members of the audience. Thomas Nagel, a philosopher visiting Rockefeller, was so impressed by Griffin's studies of bats that he wrote a paper entitled "What Is It Like to Be a Bat?" Instead of being about the thoughts and feelings of bats, Nagel's main point was that science cannot address questions that take the form "What is it like to be an X?" because science relies on objective criteria, and such questions are patently about subjective experiences, what philosophers call *qualia*. Although we can describe what Griffin's bat did when it flew around the Rockefeller dining hall, we can't say what it was like for the bat to hear the incoming echos. Similarly, we can't know what it was like to be Griffin seeing the bat, even if he tells us that it was an awe-inspiring experience. Because Griffin is a human, we know that he perceives the world as we do. He does, after all, have the same kind of brain and eyes that you and I have. But we don't know what Griffin feels when he sees a bat because these are his unique experiences.

I don't believe we will ever know what it is like, exactly, to be a bat, a bird, or a bonobo. We will never know, exactly, what the female spider monkey felt when she embraced me, looked into my eyes, and cooed. Nor will anyone ever know what it was like for me to be embraced by the spider monkey, even though my description of the encounter provides some insight into this experience. However, I also don't believe that we should be chasing this specific problem. Rather, we should use objective scientific methods to understand what animals think about and what they feel, even if we can obtain only an approximation of what is happening in their minds. The spider monkey put her arms around me in particular, not any human walking by her cage. When she put her arms around me, she did so gently. Each action appeared friendly rather than aggressive. When her mate approached, she swatted him. She thus made a distinction between individuals, cooing to one and swatting the other. These observations provide a starting point for understanding what it is like to be a female spider monkey. We will never be able to fully understand her feelings of affection, but nor can we ever say that my feelings of affection are exactly like any other human being's. I can't even say whether my own feelings of affection an hour ago are exactly the same as those I may be experiencing now. For all animals, the best we can do is to use objective criteria to show that some species share the same basic feelings and thoughts, and some species do not.

The only way to understand what animals think and feel is to explore how their minds have been designed to solve specific social and ecological problems. The same is true of the human mind. Some problems are common to all animals. As a result, we find that all animals are equipped with a universal toolkit, a set of mental abilities for acquiring knowledge about objects, number, and space. Although humans will navigate to a restaurant to eat French cuisine and then leave a 20 percent tip, while honeybees will navigate to a field of flowers and then return to the hive to waggle out the distance and direction to food, the underlying mental tools are generally similar. All animals can navigate by means of dead reckoning, and use a set of core principles to recognize objects and enumerate them. Specializations beyond these universal traits arise when unique problems emerge that demand unique solutions. Highly social animals have evolved the mental tools to watch others and learn. Specializations do not make one species "smarter" than the other, but they do make each species wonderfully different from the others. If the notion of intelligence has any role to play in the study of animal minds, it is in terms of how each species solves the problem of making a living. In the struggle to survive, nature is the only arbiter of intelligence. The survivors are smart enough to carry on living, while those that became extinct were not.

We share the planet with thinking animals. Each species, with its uniquely sculpted mind, endowed by nature and shaped by evolution, is capable of meeting the most fundamental challenges that the physical and psychological world presents. Although the human mind leaves a characteristically different imprint on the planet, we are certainly not alone in this process.

NOTES

CHAPTER 1

1 The psychologist George Miller coined the term "informavore" to describe the voracity with which we seem to acquire and seek out information. Animal curiosities: Fowler, 1970; Gallistel, 1990; Glickman and Sroges, 1966; Godin and Davis, 1995; Masson, 1997; McGrew, 1992; Thomas, 1993.
2 Different causal explanations: Alcock and Sherman, 1994; Curio, 1994; Hauser, 1996; Tinbergen, 1963.
3 Dogs on the Freudian couch: Dodman, 1996.
4 Obsessive-compulsive disorder in humans: Rappoport, 1989.
5 Standard equipment of the brain: Alman, 1998; Deacon, 1997; Hodos and Campbell, 1990.
6 Putnam, 1981.
7 Twain, 1897.
8 Le Carre, 1996, pp. 108–9.
9 I feel, therefore I think clearly: Brothers, 1997; Damasio, 1994; LeDoux, 1996.
10 Firefly femmes fatale: Eisner et al., 1997; Lloyd, 1984; Lloyd, 1986.
11 The ultimate copycats: Burghardt, 1991; McGregor and Krebs, 1984; Owen, 1980; Rowe, Coss and Owings, 1986; Wickler, 1968.

CHAPTER 2

1 Chimpanzee sandals and seats: Alp, 1997.
2 Building blocks of object knowledge: Baillargeon, 1995; Carey, 1995; Carey and Spelke, 1994; Hauser, 1998; Hood et al., 1999; Leslie, 1994; Spelke, 1991; Spelke, 1994; Spelke, Vishton and von Hofsten, 1995.
3 Nature-nurture battles: Chomsky, 1986; Chomsky, 1988; Cosmides and Tooby, 1994; Elman et al., 1996; Piaget, 1952; Pinker, 1994; Pinker, 1997; Spelke, 1994.

4 Animals and children as theorists: Carey, 1985; Carey and Spelke, 1994; Gopnik and Meltzoff, 1997; Hauser, 1998.

5 Magic tricks: Baillargeon, 1995; Hauser, 1998; Spelke, 1985.

6 Out of sight is sometimes out of mind, and sometimes in the mind: Baillargeon, 1995; Baillargeon and DeVos, 1991; Baillargeon, Spelke, and Wasserman, 1985; Diamond, 1988; Hauser, 1998; Karmiloff-Smith, 1992.

7 Under which shell is the pea hidden? Tinkelpaugh, 1928; Tinkelpaugh, 1932.

8 Monkey brains and the division of labor: Van Essen and DeYoe, 1995; Zeki, 1993.

9 How the brain solves the shell and pea trick: Watanabe, 1996.

10 Animals keep it in mind: Dore and Dumas, 1987; Gagnon and Dore, 1992; Natale et al, 1986; Pepperberg and Funk, 1990.

11 Of billiard balls and other causal interactions: Baillargeon, 1995; Heider and Simmel, 1944; Hume, 1748/1955; Kotovsky and Baillargeon, 1998; Leslie, 1982; Leslie, 1994; Leslie and Keeble, 1987; Michotte, 1962.

12 Pull the towel. Win a toy! Willatts, 1984; Willatts, 1990.

13 Neat solutions: Beck, 1980; Gould and Gould, 1995; Griffin, 1992; Hauser, 1996; Russon, 1997.

14 Stone tools: Boesch, 1991; Boesch and Boesch, 1992; Matsuzawa, 1996; Matsuzawa and Yamakoshi, 1996; McGrew, 1992.

15 Chimpanzee hammers and anvils: Boesch, 1991; Boesch and Boesch, 1984; Boesch and Boesch, 1992; Matsuzawa, 1994; Matsuzawa and Yamakoshi, 1996.

16 Crafty crows: Hunt, 1996.

17 Cunning capuchins: Visalberghi and Fragaszy, 1991; Visalberghi and Fragaszy, Savage-Rumbaugh, 1995; Visalberghi and Limongelli, 1994; Visalberghi and Limongelli, 1996.

18 Representation of "artifact" in tamarins: Hauser, 1997.

19 Signs of design: Pinker, 1997.

20 An archaeologist's perspective: Mithen, 1996.

21 Taking the design stance: Dennett, 1987; Dennett, 1996; Keil, 1994.

22 Intentions and artifacts: Bloom, 1996.

CHAPTER 3

1 T. Dantzig, 1954, "Number," cited on p. 6.

2 The mental calculus of foraging: Giraldeau, 1997; Stephens and Krebs, 1986.

3 Cooperation, coalitions, and alliances: de Waal, 1982; de Waal, 1989; de Waal, 1996; Dugatkin, 1997; Harcourt and de Waal, 1992; Wrangham and Peterson, 1996.

4 King of beasts: Grinnell, Packer, and Pusey, 1995; Grinnell and McComb, 1996; McComb et al., 1993; McComb, Packer and Pusey, 1994; Packer and Pusey, 1983; Packer and Ruttan, 1988.

5 Gastronomy and chimpanzee sounds: Clark and Wrangham, 1993; Goodall, 1986; Hauser et al., 1993; Hauser and Wrangham, 1987; Wrangham, 1977.

6 Subitization: Trick and Pylyshyn, 1994.

7 Core principles of the number system: Gallistel, 1990; Gellman and Gallistel, 1986; Karmiloff-Smith, 1992.

8 Counting up and down: Karmiloff-Smith, 1992; Spelke, Phillips and Woodward, 1995; Starkey and Cooper, 1980; Wynn, 1990; Wynn, 1995.

9 Mickey Mouse math: Wynn, 1992.
10 A clever horse named Hans: Pfungst, 1911; Sebeok and Rosenthal, 1981.
11 Rats and pigeons are numerical savants: Boysen and Capaldi, 1993; Davis and Perusse, 1988; Gallistel, 1990.
12 Animals with human tutors: Gardner and Gardner, 1969; Gardner, Gardner, & Van Cantfort, 1989; Herman, Pack, and Palmer, 1993; Herman, Richards, and Wolz, 1984; Matsuzawa, 1996; Pepperberg, 1991; Premack, 1986; Savage-Rumbaugh, 1986; Schusterman, et al., 1993; Terrace, 1979.
13 Tutored animals, concept formation, and number: Boysen, 1996; Boysen and Capaldi, 1993; Matsuzawa, 1985; Pepperberg, 1987; Pepperberg, 1994; Premack and Premack, 1994; Premack and Woodruff, 1978; Rumbaugh and Washburn, 1993; Washburn and Rumbaugh, 1991.
14 Inhibiting selfish impulses: Boysen, 1996; Boysen and Berntson, 1995.
15 Magic and monkey number: Hauser, 1997; Hauser and Carey, 1998; Uller, 1997.
16 Number principles: Dehaene, 1997; Gallistel, 1990; Hauser, 1997.
17 Counting your kids: Clutton-Brock, 1992.
18 Gossiping in numbers: Barton and Dunbar, 1997; Dunbar, 1996.

CHAPTER 4

1 Amazing pet stories: Steiger and Steiger, 1992.
2 Sex, ecology, and the brain: Healy, 1996; Sherry, 1997; Sherry and Hampson, 1997; Sherry, Jacobs, and Gaulin, 1992.
3 Ants and the dunes: Wehner, 1997; Wehner and Menzel, 1990; Wehner, Michel, and Antonsen, 1996.
4 Thinking about shortcuts: Gallistel, 1990; Poucet, 1993; Thinus-Blanc, 1995.
5 Sun compasses: Able, 1991; Gallistel, 1990; Santchi, 1913.
6 Internal tickers: Gallistel, 1990; Gwinner, 1996; Shettleworth, 1998.
7 Sun birds: Chappell and Guilford, 1995; Duff, Brownlie, and Sherry, 1999; Emlen, 1975; Wiltschko and Balda, 1989; Gould and Gould, 1988.
8 Expert landmark use among food-storing birds: Duff, Brownlie, and Sherry, in press; Kamil and Jones, 1997; Sherry, Jacobs, and Gaulin, 1992; Sherry et al., 1989.
9 Accurate travel by smell, touch, and magnetic force fields: Able, 1996; Dittman and Quinn, 1996; Landau, Spelke, and Gleitman, 1984; Walker et al., 1997; Wallraff, 1996; Wiltschko and Wiltschko, 1996.
10 Short cuts and GPS systems: Bennett, 1996; Gallistel, 1990; Gallistel, 1998; Gallistel and Cramer, 1996; O'Keefe and Nadel, 1978; Thinus-Blanc, 1995; Tolman, 1955.
11 Skeptic bees with maps in the mind or well-worn routes? Dyer, 1996; Gould and Gould, 1995; Menzel et al., 1996; Wehner, Michel, and Antonsen, 1996.
12 Apes in space: Boesch and Boesch, 1984; Gallistel and Cramer, 1996; Menzel, 1973; Savage-Rumbaugh, 1986; Savage-Rumbaugh and Lewin, 1996; Savage-Rumbaugh et al., 1986.
13 Mapping from miniatures to real life: DeLoache and Brown, 1983; Premack and Premack, 1983.
14 Cognitive maps denied: Bennett, 1996; Wehner and Menzel, 1990.
15 A curiosity for landmarks: Biegler and Morris, 1996; Collett, 1996; Gallistel, 1990; Joubert and Vauclair, 1985; Judd and Collett, 1998; Menzel and Menzel, 1979.

16 Lost in space: Cheng, 1986; Hermer and Spelke, 1996; Hermer and Spelke, 1994; Margules and Gallistel, 1988.

17 Cruising for rodent mates: Gaulin and FitzGerald, 1986; Gaulin and FitzGerald, 1989; Jacobs and Spencer, 1994.

18 Comparative brain stories: Healy, 1996; Sherry, 1997; Sherry and Hampson, Jacobs, 1997; Sherry, Jacobs, and Gaulin, 1992.

19 The brains behind food-storing birds: Balda and Kamil, 1989; Basil et al., 1996; Clayton, 1995; Clayton and Krebs, 1994; Healy, 1996; Krebs et al., 1989; Sherry, 1997; Sherry and Vaccarino, 1989; Sherry et al., 1989.

20 Domain-general or -specific? Clayton, 1995; Sherry, 1997; Sherry et al., 1993.

CHAPTER 5

1 I? Me? Who? Bermudez, Marcel and Eilan, 1995.

2 A chick's-eye view of mother hen: Bateson, 1966; Horn, 1991; Johnson and Morton, 1991; Lorenz, 1937.

3 Kin signatures: Bourke, 1997; Dawkins, 1976; Grosberg, 1988; Maynard Smith, 1964; Sherman, Reeve, and Pfennig, 1997.

4 They're playing my song: Beecher, 1982; Margoliash, 1983; Margoliash, 1987; Marler, 1970; Marler and Tamura, 1962; McGregor, 1992.

5 Whose face? Damasio, 1994; Damasio, Damasio, and Tranel, 1990a; Damasio, Tranel, and Damasio, 1990b; Tovee and Cohen-Tovee, 1993; Tranel and Damasio, 1985; Tranel and Damasio, 1993; Tranel, Damasio, and Damasio, 1988; Young, 1992.

6 Invasion of the body snatchers: Ellis and Young, 1990; Hirstein and Ramachandran, 1997.

7 Darwin, 1871, p. 62.

8 Darwin, 1872, p. 140.

9 Monkeys and apes in the mirror: Gallup and Povinelli, 1993; Gallup, 1970; Gallup, 1987; Gallup, 1991; Mitchell, 1993; Povinelli, 1987; Povinelli, 1993; Povinelli et al., 1993.

10 Objects in the mirror: Anderson, 1984; Brown, McDowell, and Robinson, 1965; Moody, 1975; Pepperberg et al., 1995; Povinelli, 1989.

11 Revenge of the pigeons: Epstein, Lanza, and Skinner, 1981.

12 Mirrored variations: Povinelli et al., 1993; Swartz and Evans, 1991; Swartz and Evans, 1997.

13 Chimps on TV: Menzel, Savage-Rumbaugh, and Lawson, 1985.

14 Punk monkeys: Hauser et al., 1995.

15 Damaged brain reveals the stuff of self: Ramachandran and Blakeslee, 1998.

16 Clones, twins, and individuality: Bouchard, 1997.

17 Two-headed snake: Burghardt, 1991.

CHAPTER 6

1 With a little help from their friends: Cristol et al., 1998; Dugatkin, 1997; Ingmanson, 1996; Meltzoff and Moore, 1995; Nishida, 1987; Rawls, Fiorelli, and Gish, 1985; Tayler and Saayman, 1973.

2 Equipment for vocal learning in birds and babies: Catchpole and Slater, 1995;

Hauser, 1996; Kroodsma and Miller, 1996; Locke, 1993; Marler, 1970; Marler, 1997; Pinker, 1994; Snowdon and Hausberger, 1997.

3 The art of reverse engineering: Cosmides and Tooby, 1992; Cosmides and Tooby, 1994; Daly, 1997; Pinker, 1997.

4 Instincts to learn: Marler, 1989; Pinker, 1994.

5 Song cultures: Baker and Jenkins, 1987; Payne, Payne, and Doehlert, 1988.

6 Bird brains: Alvarez-Buylla, Kirn, and Nottebohm, 1990; Brenowitz and Kroodsma, 1996; Konishi, 1985; Nottebohm, Kasparian, and Pandazis, 1981.

7 Plasticity and the case of the missing limb: Ramachandran and Blakeslee, 1998.

8 Mechanical constraints on the voice: Fitch, 1997; Fitch and Hauser, 1995; Hauser, 1996; Lieberman, 1984.

9 *Homo imitans* and *sapiens:* Meltzoff, 1988; Meltzoff, 1996; Meltzoff and Moore, 1995.

10 The black rats of Israel: Terkel, 1996.

11 Lab rats on the information highway: Galef, 1996.

12 Social learning everywhere: Fiorito and Scotto, 1992; Heyes, 1996; Warner, 1988.

13 Date copying in animals: Clutton-Brock and McComb, 1993; Dugatkin, 1992; Gibson, Bradbury, and Vehrencamp, 1991.

14 Guppy sex: Dugatkin, 1997; Houde and Endler, 1990.

15 Kinds of imitation: Byrne, 1995; Russon, 1997; Tomasello and Call, 1997; Whiten and Custance, 1996.

16 Orangutan rehabilitation and the art of imitation: Russon and Galdikas, 1993; Russon and Galdikas, 1995.

17 Swift acquisition, imitation, and birds with a taste for cream: Boyd and Richerson, 1985; Fisher and Hinde, 1949; Laland, Richerson, and Boyd, 1996; Sherry and Galef, 1984; Sherry and Galef, 1990.

18 Do as I do: Custance, Whiten, and Bard, 1995; Hayes and Hayes, 1952; Tomasello, 1996; Tomasello, Kruger, and Ratner, 1993; Whiten and Custance, 1996.

19 Myths of monkey imitation: Cheney and Seyfarth, 1990; Fragaszy and Visalberghi, 1996; Hauser, 1996; Snowdon and Hausberger, 1997; Tomasello, 1996; Visalberghi and Fragaszy, 1991.

20 Parrot parrot-ing: Moore, 1992; Moore, 1996.

21 Rats do it: Heyes, 1996; Heyes and Dawson, 1990; Heyes et al., 1994.

22 Imitation, the next generation: Byrne, 1995; Galef, 1988; Galef, 1996; Heyes, 1996; Heyes et al., 1994; Tomasello, 1996; Tomasello and Call, 1997; Whiten and Custance, 1996.

23 Thinking about teaching: Boesch, 1996; Caro and Hauser, 1992; Parker, 1996; Parker and Russon, 1996.

24 Chimp teaching: Boesch, 1991; Boesch, 1996; Matsuzawa, in press.

25 Cat teaching: Caro, 1980a; Caro, 1980b; Caro, 1994; Caro and Hauser, 1992.

CHAPTER 7

1 Birds that cry wolf: Munn, 1986a; Munn, 1986b.

2 Handicaps and honesty: Collins, 1993; Grafen, 1990; Hauser, 1996; Johnstone and Grafen, 1993; Reeve, 1997; Viljugrein, 1997; Zahavi, 1975; Zahavi and Zahavi, 1997.

3 Symmetry rules: Møller, 1993; Møller, 1994; Møller and Swaddle, 1997; Thornhill and Gangestad, 1998; Van Valen, 1962.

4 Bluffing shrimp: Caldwell, 1986; Caldwell and Dingle, 1975; Steger and Caldwell, 1983.

5 Sly roosters: Evans and Marler, 1995; Margoliash, 1983; Marler, Dufty, and Pickert, 1986a; Marler, Dufty, and Pickert, 1986b; Marler, Karakashian, and Gyger, 1991.

6 Silent cheaters: Hauser, 1997; Hauser and Marler, 1993a; Hauser and Marler, 1993b.

7 Purposive chimps play follow the leader: Menzel and Halperin, 1975.

8 Skeptics: Godfray, 1991; Gould and Gould, 1995; Grafen, 1990; Hauser, 1993; Johnstone, 1997; Kilner, 1997.

9 Primate lies: Byrne and Whiten, 1990; Savage-Rumbaugh, and Lewin, 1996.

10 Bird feigns injury! Predators believe: Ristau, 1991a; Ristau, 1991b.

11 Mind-reading humans: Baron-Cohen, 1995; Baron-Cohen, Tager-Flusberg, and Cohen, 1993; Carruthers and Smith, 1996; Frye and Moore, 1991; Lewis and Mitchell, 1994; Perner, 1991; Wellman, 1990; Premack and Woodruff, 1978.

12 Objects and animacy: Premack and Premack, 1995; Premack and Premack, 1997; Gergely et al., 1995.

13 The infant's path to becoming mind-literate: Baron-Cohen, 1995; Butterworth, 1991; Clements and Perner, 1994; Dunn, 1994; Gergely et al., 1995; Gopnik and Meltzoff, 1997; Hala and Chandler, 1997; Leslie, 1994; Lewis and Mitchell, 1994; Perner, 1991; Slaughter and Gopnik, 1997; Wellman, 1990; Wimmer and Perner, 1983.

14 Chimpanzee knowledge on trial: Heyes, 1998; Premack and Premack, in press.

15 Ignorant about ignorance: Cheney and Seyfarth, 1988; Cheney and Seyfarth, 1990; Gagliardi et al., 1995; Heyes, 1993; Heyes, 1998; Povinelli, Nelson, and Boysen, 1990; Povinelli, Nelson, and Boysen, 1992a; Povinelli, Parks, and Novak, 1991; Povinelli, Parks, and Novak, 1992b.

16 Neurophysiological specificity for eyes and heads: Perrett et al., 1990; Perrett et al., 1991; Tovee and Cohen-Tovee, 1993.

CHAPTER 8

1 Lions and chickens, language and thought: Premack, 1986; Wittgenstein, 1958.

2 From gavagai to detached rabbit parts: Fodor, 1994; Premack, 1986; Quine, 1960; Quine, 1973.

3 What is the meaning of ham sandwich? Fauconnier, 1994.

4 What's a word? Bloom, 1994; Carey, 1982; Jackendoff, 1983; Lakoff, 1987; Macnamara, 1982; Markman, 1990; Markman and Hutchinson, 1984; Premack, 1990.

5 Vocal signatures: Beecher, 1982; Gouzoules and Gouzoules, 1990; Hauser, 1992; Johnstone, 1997; McGowan and Reiss, 1997; Rendall, Rodman, and Edmond, 1996; Snowdon and Hausberger, 1997; Tyack, 1993.

6 Cry babies: Colton and Steinschneider, 1980; Dawkins and Krebs, 1978; Evans, 1994; Godfray, 1991; Green and Gustafson, 1983; Gustafson, Green, and Tomic, 1984; Gustafson, Green, and Cleland, 1994; Hauser, 1993; Kilner, 1997; Lester and Boukydis, 1992; Lieberman, 1985; Trivers, 1972; Trivers, 1974; Zeskind et al., 1985.

7 Word learning problems: Bloom and Markson, 1998; Miller, 1996; Pinker, 1994; Redington and Chater, 1997; Saffran, Aslin, and Newport, 1996; Tomasello, 1988.

8　Hearing out of our range: Garstang et al., 1995; Griffin, 1958; Langbauer et al., 1991; Simmons, 1990.

9　Animal communication, *sans* language: Bradbury and Vehrencamp, 1998; Sebeok, 1977.

10　Pioneers on primate communication: Green, 1975; Marler, 1976; Struhsaker, 1967.

11　Honeybees dance up a storm: Gould, 1990; Gould and Gould, 1988; Michelsen, 1992; Michelsen et al., 1992; von Frisch, 1967.

12　Reference, reference everywhere: Cheney and Seyfarth, 1990; Evans and Marler, 1995; Hauser, 1996; Hauser and Wrangham, 1987; Macedonia, 1991; Marler, 1976; Marler, Dufty, and Pickert, 1986a; Marler, Dufty, and Pickert, 1986b.

13　Acoustic and referential similarity: Cheney and Seyfarth, 1988; Hauser, 1998; Zuberbuhler, Noe, and Seyfarth, 1997.

14　Do animals have words? Cheney and Seyfarth, 1990; Deacon, 1997; Hauser, 1996; Pinker, 1994.

15　Taking the intentional stance: Cheney; and Seyfarth, 1990; Dennett, 1987; Dennett, 1983.

16　Rules of order: Green and Marler, 1979; Isaac and Marler, 1963; Marler, 1977; Marler et al.,1988; Matsuzawa, 1996.

17　Grammatical particles and parsing: Abler, 1989; Abler, 1997; Hirsh-Patek and Golinkoff, 1991; Pinker, 1994; Studdert-Kennedy, 1998.

18　Whale tales: Payne, Tyack, and Payne, 1984; Payne and Payne, 1985.

19　Sparrow song: Marler and Peters, 1989; Marler and Pickert, 1984.

20　Grammatical chickadees? Hailman and Ficken, 1987; Hailman, Ficken, and Ficken, 1985; Hailman, Ficken, and Ficken, 1987.

21　Monkey grammarians? Cleveland and Snowdon, 1981; Robinson, 1979; Robinson, 1984.

22　Feral children: Candland, 1995.

23　Chimpanzees can't speak: Crelin, 1987; Hauser, 1996; Lieberman, 1984; Negus, 1949.

24　Animal language data, history, and wars: Fouts and Mills, 1997; Gardner, Gardner, and Van Cantfort, 1989; Herman, Pack, and Palmer, 1993; Patterson, 1987; Pepperberg, 1991; Premack, 1986; Savage-Rumbaugh, 1986; Savage-Rumbaugh and Lewin, 1996; Schusterman et al., 1993; Wallman, 1992.

25　Premack and Premack, 1983.

26　Language evolution: Hurford, Studdert-Kennedy and Knight, 1998; Jackendoff, in press.

CHAPTER 9

1　Other reductionists: de Waal, 1996; Premack and Premack, 1995; Ridley, 1996; Wright, 1994.

2　Pagels, 1988.

3　Emotional wisdom: Gibbard, 1990.

4　Parental guidance important: Baldwin and Moses, 1996; Campos and Stenberg, 1981.

5　Human infants and the living-nonliving distinction: Baillargeon, 1995; Leslie, 1982; Leslie, 1994, Leslie and Keeble, 1987; Spelke, Phillips, and Woodward, 1995; Woodward, 1998.

6 Seeing emotions in motion: Heider and Simmel, 1944; Michotte, 1962.

7 Intentional stance: Dennett, 1987.

8 The biology of emotion: Damasio, 1994; Davidson, 1995; Ekman, Levenson, and Friesen, 1983; LeDoux, 1996.

9 Monkey expectations about object location: Hauser, 1998.

10 Ethical experiments or experiments on ethics? Church, 1959; Lavery and Foley, 1963; Masserman, Wechkin, and William, 1964; Miller, 1967; Miller, Banks, and Kuwahara, 1966; Miller, Banks, and Ogawa, 1962; Miller, Banks, and Ogawa, 1963; Wechkin, Masserman, and Terris, 1964.

11 Primate smiles and grins: Chevalier-Skolnikoff, 1973; Fridlund, 1994; Hauser, 1996; Huber, 1931.

12 The economics of animal needs and desires: Dawkins, 1983; Dawkins, 1990; Mason, Garner, and McFarland, 1998.

13 Metaphysical and moral aspects of death and dying: Fischer, 1993.

14 How we acquire a theory of biology: Carey, 1985; Hatano and Inagaki, 1995; Keil, 1994.

15 Animal response to death: Allen and Hauser, 1991; de Waal, 1996; Masson and McCarthy, 1995.

16 Death in elephants: Moss, 1987, p. 73.

17 Rational passions: Damasio, 1994; Rottschaefer, 1998.

18 Boxed in: Diamond, 1988.

19 Front end heavy: Deacon, 1997; Diamond, 1988; Goldman-Rakic, 1987.

20 When tamarins are boxed in: Santos, Ericson, and Hauser, in press.

21 Greedy kids: Russel et al., 1991.

22 Inhibition or reinforcement? Silberberg and Fujita, 1996.

23 Scientific revolutions and paradigm shifts: Kuhn, 1970; Lakatos, 1978.

24 Conceptual change in evolution and over development: Carey, 1985; Cosmides and Tooby, 1994; Gopnik and Meltzoff, 1997; Keil, 1994; Pinker, 1997.

25 Human infants and adult monkeys pass Newton's gravity test: Hood, 1995; Hood et al., 1999.

26 How evolutionary intuitions invade unsuspecting disciplines: Anders, 1994; Barkow, Cosmides, and Tooby, 1992; Binmore, 1992; Daly and Wilson, 1988; Dennett, 1995; Dissanayake, 1992; Posner, 1992.

27 The inspiration from selfish genes: Dawkins, 1976; Dawkins, 1986; Hamilton, 1964a; Hamilton, 1964b; Hamilton, 1971; Maynard Smith, 1964; Maynard Smith, 1965; Maynard Smith, 1974; Trivers, 1971; Trivers, 1972; Trivers, 1974; Williams, 1966.

28 The evolution of reciprocal altruism: Dugatkin, 1997; Ridley, 1996; Sober and Wilson, 1998.

29 The science of cooperative games: Axelrod, 1984; Boyd, 1992; Dugatkin, 1997.

30 Coalitions, cooperation, and reciprocal altruism: Connor, Smolker, and Richards, 1992; de Waal, 1989a, de Waal, 1996; Harcourt and de Waal, 1992; Heinsohn and Packer, 1995; Hemelrijk, 1990.

31 Big brains and social complexity: Barton and Dunbar, 1997; Dunbar, 1996; Harcourt, 1988.

32 Of vampire bats and guppies—born reciprocal altruists: Dugatkin, 1997; Wilkinson, 1984.

33 Human games of fairness and deception: Binmore, 1994; Binmore, 1998; Cosmides and Tooby, 1992.

34 Subordinates with special skills are treated like royalty: Stammbach, 1988.

35 Value, ownership, respect, and normative responses: Gibbard, 1990; Kummer, 1978; Kummer and Cords, 1991.

36 When other animals are perceived as property: Bachmann and Kummer, 1980; Beach, Buehler, and Dunbar, 1982; Kummer, Gotz, and Angst, 1974; Sigg and Falett, 1985.

37 Punishment in the animal kingdom: Clutton-Brock and Parker, 1995; Hauser, 1997.

38 Friendly, reconciling, peaceful primates: Cords and Aureli, 1993; Cords and Thurn-heer, 1993; de Waal, 1982, de Waal, 1989b, de Waal, 1996; de Waal and Johanowicz, 1993; de Waal and van Roosmalen, 1979; de Waal and Yoshihara, 1983; Silk, in press.

39 Debates about animal rights: Cavalieri and Singer, 1993; Clark, 1997; Petrinovich, 1995; Petrinovich, 1999; Rollin, 1989.

REFERENCES

CHAPTER 1

Alcock, J., and P. Sherman. (1994). The utility of the proximate-ultimate dichotomy in ethology. *Ethology*, 96, 58–62.

Alman, J. (1998). *Evolving Brains*. New York: W. H. Freeman.

Au, W. L. (1993). *The Sonar of Dolphins*. Berlin: Springer-Verlag.

Brothers, L. (1997). *Friday's Footprints*. Oxford: Oxford University Press.

Burghardt, G. M. (1991). Cognitive ethology and critical anthropomorphism: A snake with two heads and hognose snakes that play dead. In C. A. Ristau (Ed.), *Cognitive Ethology: The Minds of Other Animals* (pp. 53–91). Hillsdale: Lawrence Erlbaum.

Curio, E. (1994). Causal and functional questions: How are they linked? *Animal Behaviour*, 47, 999–1021.

Damasio, A. (1994). *Descartes' Error: Emotion, Reason and the Human Brain*. New York: Putnam.

Deacon, T. W. (1997). *The Symbolic Species: The Co-evolution of Language and the Brain*. New York: Norton.

Dodman, N. (1996). *The Dog Who Loved Too Much*. New York: Bantam Books.

Eisner, T., M. A. Goetz, D. E. Hill, S. R. Smedley, and J. Meinwald (1997). Firefly "femmes fatales" acquire defensive steroids (lucibufagins) from their firefly prey. *Proceedings of the National Academy of Sciences*, 94, 9723–9728.

Fowler, H. (1970). *Curiosity and Exploratory Behavior*. New York: Macmillan.

Gallistel, C. R. (1990). *The Organization of Learning*. Cambridge: MIT Press.

Glickman, S. E., and R.W. Sroges. (1966). Curiosity in zoo animals. *Behaviour*, 26, 151–88.

Godin, J. G., and S. A. Davis. (1995). Who dares, benefits: Predator approach behaviour in the guppy (Poecilia reticulata) deters predator pursuit. *Proceedings of the Royal Society, London*, 259, 193–200.

Griffin, D. R. (1958). *Listening in the Dark*. New Haven: Yale University Press.

Hauser, M. D. (1996). *The Evolution of Communication*. Cambridge: MIT Press.

Hodos, W., and C. B. G. Campbell (1990). Evolutionary scales and comparative studies of animal cognition. In R. P. Kesner and D. S. Olton (Eds.) *Neurology of Comparative Cognition* (pp. 1–20). Hillsdale, N.J.: Lawrence Erlbaum Assoc.

Hoy, R. (1992). The evolution of hearing in insects as an adaptation to predation from bats. In D. B. Webster, R. F. Fay, and A. N. Popper (Eds.), *The Evolutionary Biology of Hearing* (pp. 115–30). New York: Springer-Verlag.

Le Carré, J. (1997). *The Tailor of Panama*. Ballantine: New York.

LeDoux, J. (1996). *The Emotional Brain: The Mysterious Underpinnings of Emotional Life*. New York: Simon and Schuster.

Lloyd, J. E. (1984). On deception, a way of all flesh, and firefly signalling and systematics. In R. Dawkins and M. Ridley (Eds.), *Oxford Surveys in Evolutionary Biology, vol 1* (pp. 48–54), New York: Oxford University Press.

———. (1986). Firefly communication and deception: "Oh, what a tangled web." In R. W. Mitchell and N. S. Thompson (Eds.), *Deception: Perspectives on Human and Nonhuman Deceit* (pp. 113–128). Albany: State University of New York Press.

Masson, J. M. (1997). *Dogs Never Lie about Love*. New York: Crown.

McGregor, P. K., and J. R. Krebs. (1984). Song learning and deceptive mimicry. *Animal Behaviour, 32*, 280–87.

McGrew, W. C. (1992). *Chimpanzee Material Culture*. Cambridge: Cambridge University Press.

Owen, D. (1980). *Camouflage and Mimicry*. Chicago: University of Chicago Press.

Pinker, S. (1997). *How the Mind Works*. New York: Norton.

Putnam, H. (1981). *Reason, Truth and History*. Cambridge: Cambridge University Press.

Rappoport, J. L. (1989). *The Boy Who Couldn't Stop Washing*. New York: Penguin Books.

Rowe, M. P., R. G. Coss, and D. H. Owings. (1986). Rattlesnake rattles and burrowing old hisses: A case of acoustic Batesian mimicry. *Ethology, 72*, 53–71.

Simmons, J. (1990). A view of the world through the bat's ear: The formation of acoustic images in echolocation. *Cognition, 33*, 155–99.

Thomas, E. M. (1993). *The Hidden Life of Dogs*. New York: Houghton Mifflin.

Tinbergen, N. (1963). On aims and methods in ethology. *Zeitschrift für Tierpsychologie, 20*, 410–33.

Twain, M. (1897). *Pudd'nhead Wilson's New Calendar*. London: Rover.

Wickler, W. (1968). *Mimicry in Plants and Animals*. New York: McGraw-Hill.

CHAPTER 2

Alp, R. (1997). "Stepping-sticks" and "Seat-sticks": New types of tools used by wild chimpanzees (*Pan troglodytes*) in Sierra Leone. *American Journal of Primatology, 41*, 45–52.

Baillargeon, R. (1995). Physical reasoning in infancy. In M. Gazzaniga (Ed.), *The Cognitive Neurosciences* (pp. 181–204). Cambridge: MIT Press.

Baillargeon, R., and J. DeVos. (1991). Object permanence in young infants: Further evidence. *Child Development, 62*, 1227–1246.

Baillargeon, R., E. Spelke, and S. Wasserman. (1985). Object permanence in five month old infants. *Cognition, 20*, 191–208.

Basolo, A. L. (1990). Female preference predates the evolution of the sword in sword-tails. *Science*, *250*, 808–10.

Beck, B. B. (1980). *Animal Tool Behavior*. New York: Garland.

Bloom, P. (1996). Intention, history, and artifact concepts. *Cognition*, *60*, 1–29.

Boesch, C. (1991). Teaching among wild chimpanzees. *Animal Behaviour*, *41*, 530–32.

Boesch, C., and H. Boesch. (1984). Mental map in wild chimpanzees: An analysis of hammer transports for nut cracking. *Primates*, *25*, 160–70.

———. (1992). Transmission aspects of tool use in wild chimpanzees. In T. Ingold and K. R. Gibson (Eds.), *Tools, Language and Intelligence: Evolutionary Implications* (pp. 171–83). Oxford: Oxford University Press.

Carey, S. (1985). *Conceptual Change in Childhood*. Cambridge: MIT Press.

———. (1995). On the origin of causal understanding. In D. Sperber, D. Premack, and A. Premack (Eds.), *Causal Cognition* (pp. 113–46). Oxford: Clarendon Press.

Carey, S., and E. S. Spelke. (1994). Domain-specific knowledge and conceptual change. In L. A. Hirschfeld and S. A. Gelman (Eds.), *Mapping the Mind: Domain-Specificity in Cognition and Culture* (pp. 169–201). Cambridge: Cambridge University Press.

Cheney, D. L., and R. M. Seyfarth. (1990). *How Monkeys See the World: Inside the Mind of Another Species*. Chicago: University of Chicago Press.

Chomsky, N. (1986). *Knowledge of Language: Its Nature, Origin, and Use*. New York: Praeger.

———. (1988). *Language and Problems of Knowledge*. Cambridge: MIT Press.

Cosmides, L., and J. Tooby. (1994). Beyond intuition and instinct blindness: Toward an evolutionarily rigorous cognitive science. *Cognition*, *50*, 41–77.

Dennett, D. (1987). *The Intentional Stance*. Cambridge: MIT Press.

———. (1996). *Kinds of Minds*. New York: Basic Books.

Diamond, A. (1988). Differences between adult and infant cognition: Is the crucial variable presence or absence of language? In L. Weiskrantz (Ed.), *Thought without Language* (pp. 337–70). Oxford: Clarendon Press.

Dore, F. Y., and C. Dumas. (1987). Psychology of animal cognition: Piagetian studies. *Psychological Bulletin*, *102*, 219–33.

Elman, J., E. Bates, M. Johnson, A. Karmiloff Smith, J. Parisi, and J. Plunkett. (1996). *Rethinking Innateness*. Cambridge: MIT Press.

Gagnon, S., and F. Y. Dore. (1992). Search behavior in various breeds of adult dogs (*Canis familiaris*): Object permanence and olfactory cues. *Journal of Comparative Psychology*, *106*, 58–68.

Gopnik, A., and A. Meltzoff. (1997). *Words, Thoughts, and Theories*. Cambridge: MIT Press.

Gould, J., and C. Gould. (1995). *The Animal Mind*. New York: Scientific American.

Greenfield, P. M. (1991). Language, tools, and brain: The ontogeny and phylogeny of hierarchically organized sequential behavior. *Behavioral and Brain Sciences*, *14*, 531–95.

Griffin, D. R. (1992). *Animal Minds*. Chicago: University of Chicago Press.

Hauser, M. D. (1996). *The Evolution of Communication*. Cambridge: MIT Press.

———. (1997). Artifactual kinds and functional design features: What a primate understands without language. *Cognition*, *64*, 285–308.

———. (1998). Expectations about object motion and destination: Experiments with a nonhuman primate. *Developmental Science 1*, 31–38.

Heider, F., and M. Simmel. (1944). An experimental study of apparent behavior. *American Journal of Psychology*, 57, 243–59.

Herrnstein, R. J. (1991). Levels of categorization. In G. M. Edelman, W. E. Gall, and W. M. Cowan (Eds.), *Signal and Sense: Local and Global Order in Perceptual Maps* (pp. 385–413). New York: Wiley-Liss.

Herrnstein, R. J., D. H. Loveland, and C. Cable. (1976). Natural concepts in pigeons. *Journal of Experimental Psychology (Animal Behavior)*, 2, 285–311.

Hood, B. M., M. D. Hauser, L. Anderson, and L. Santos. (1999). Gravity biases in a nonhuman primate? *Developmental Science*, 2, 35–41.

Hume, D. (1748/1955). *Inquiry Concerning Human Understanding*. Indianapolis: Bobbs-Merrill.

Hunt, G. R. (1996). Manufacture of hook-tools by New Caledonian crows. *Nature*, 379, 249–51.

Karmiloff-Smith, A. (1992). *Beyond Modularity*. Cambridge: MIT Press.

Keil, F. C. (1994). The birth and nurturance of concepts by domains: The origins of concepts of living things. In L. A. Hirschfeld and S. A. Gelman (Eds.), *Mapping the Mind: Domain—Specificity in Cognition and Culture* (pp. 234–54). Cambridge: Cambridge University Press.

Kotovsky, L., and R. Baillargeon. (1998). Calibration-based reasoning about collision events in 11 month-old infants. *Cognition*, 67, 311–51.

Leslie, A. M. (1982). The perception of causality in infants. *Perception*, 11, 173–86.

———. (1994). ToMM, ToBy, and Agency: Core architecture and domain specificity. In L. A. Hirschfeld and S. A. Gelman (Eds.), *Mapping the Mind: Domain—Specificity in Cognition and Culture* (pp. 119–48). Cambridge: Cambridge University Press.

Leslie, A. M., and S. Keeble. (1987). Do six-month old infants perceive causality? *Cognition*, 25, 265–88.

Locke, J. (1993). *The Path to Spoken Language*. Cambridge: Harvard University Press.

Matsuzawa, T. (1994). Field experiments on use of stone tools in the wild. In R. W. Wrangham, W. C. McGrew, F. B. M. de Waal, and P. G. Heltne (Eds.), *Chimpanzee Cultures* (pp. 351–70). Cambridge: Harvard University Press.

———. (1996). Chimpanzee intelligence in nature and in captivity: Isomorphism of symbol use and tool use. In W. C. McGrew, L. F. Nishida, and T. Nishida (Eds.), *Great Ape Societies* (pp. 196–209). Cambridge: Cambridge University Press.

Matsuzawa, T., and G. Yamakoshi. (1996). Comparison of chimpanzee material culture between Bossou and Nimba, West Africa. In A. E. Russon, K. A. Bard, and S. T. Parker (Eds.), *Reaching into Thought: The Minds of the Great Apes* (pp. 211–34). Cambridge: Cambridge University Press.

McGrew, W. C. (1992). *Chimpanzee Material Culture*. Cambridge: Cambridge University Press.

Michotte, A. (1962). *The Perception of Causality*. Andover: Methuen.

Mithen, S. (1996). *A Prehistory of the Mind*. New York: Bantam Books.

Møller, A. P. (1994). *Sexual Selection and the Barn Swallow*. Oxford: Oxford University Press.

Natale, F., F. Antinucci, G. Spinozzi, and. P. Poti. (1986). Stage 6 object concept in nonhuman primate cognition: A comparison between gorilla (*Gorilla gorilla gorilla*) and Japanese macaque (*Macaca fuscata*). *Journal of Comparative Psychology*, 100, 335–39.

Pepperberg, I. M., and F. A. Funk. (1990). Object permanence in four species of psittacine birds: An African Grey parrot (*Psittacus erithacus*), an Illiger mini macaq (*Ara maracana*), a parakeet (*Melopsittacus undulatus*), and a cockatiel (*Nymphicus hollandicus*). *Animal Learning and Behavior*, *14*, 322–30.

Piaget, J. (1952). *The Origins of Intelligence in Children*. New York: International University Press.

Pinker, S. (1994). *The Language Instinct*. New York: William Morrow.

———. (1997). *How the Mind Works*. New York: Norton.

Russon, A. E. (1997). Exploiting the expertise of others. In A. Whiten and R. W. Byrne (Eds.), *Machiavellian Intelligence 2* (pp. 174–206). Cambridge: Cambridge University Press.

Spelke, E. S. (1985). Preferential looking methods as tools for the study of cognition in infancy. In G. Gottlieb and N. Krasnegor (Eds.), *Measurement of Audition and Vision in the First Year of Postnatal Life* (pp. 85–168). Norwood: Ablex.

———. (1991). Physical knowledge in infancy: Reflections on Piaget's theory. In S. Carey and R. Gelman (Eds.), *The Epigenesis of Mind: Essays on Biology and Cognition* (pp. 37–61). Hillsdale: Lawrence Erlbaum.

———. (1994). Initial knowledge: Six suggestions. *Cognition*, *50*, 431–45.

Spelke, E. S., P. Vishton, and C. von Hofsten. (1995). Object perception, object-directed action, and physical knowledge in infancy. In M. Gazzaniga (Ed.), *The Cognitive Neurosciences* (pp. 165–79). Cambridge: MIT Press.

Thompson, R. K. R. (1995). Natural and relational concepts in animals. In H. L. Roitblat and J. A. Meyer (Eds.), *Comparative Approaches to Cognitive Science* (pp. 175–224). Cambridge: MIT Press.

Tinkelpaugh, O. L. (1928). An experimental study of representative factors in monkeys. *Journal of Comparative Psychology*, *8*, 197–236.

———. (1932). Multiple delayed reaction with chimpanzees and monkeys. *Journal of Comparative Physiological Psychology*, *13*, 207–24.

Tomonaga, M. (1996). Visual perception in chimpanzees (*Pan troglodytes*): Texture segregation and perception of shape from shading. *The Emergence of Human Cognition and Language*, *3*, 89–98.

Van Essen, D. C., and E. A. DeYoe. (1995). Concurrent processing in the primate visual cortex. In M. Gazzaniga (Ed.), *The Cognitive Neurosciences* (pp. 383–400). Cambridge: MIT Press.

Visalberghi, E., and D. Fragaszy. (1991). Do monkeys ape? In S. T. Parker and K. R. Gibson (Eds.), *"Language" and Intelligence in Monkeys and Apes* (pp. 247–73), Cambridge University Press: Cambridge.

Visalberghi, E., D. M. Fragaszy, and E. S. Savage-Rumbaugh. (1995). Performance in a tool-using task by common chimpanzees (*Pan troglodytes*), bonobos (*Pan paniscus*), an orangutan (*Pongo pygmaeus*), and capuchin monkeys (*Cebus apella*). *Journal of Comparative Psychology*, *109*, 52–60.

Visalberghi, E., and L. Limongelli. (1994). Lack of comprehension of cause-effect relations in tool-using capuchin monkeys (*Cebus apella*). *Journal of Comparative Psychology 108*, 15–22.

———. (1996). Acting and understanding: Tool use revisited through the minds of capuchin monkeys. In A. E. Russon, K. A. Bard, and S. T. Parker (Eds.), *Reaching into*

Thought: The Minds of the Great Apes (pp. 57–79). Cambridge: Cambridge University Press.

Watanabe, M. (1996). Reward expectancy in primate prefrontal neurons. *Nature, 382,* 629–32.

Willatts, P. (1984). The Stage-IV infant's solution of problems requiring the use of supports. *Infant Behavior and Development, 7,* 125–34.

———. (1990). Development of problem-solving strategies in infancy. In D. F. Bjorklund (Ed.), *Children's Strategies: Contemporary Views of Cognitive Development* (pp. 143–82). Hillsdale: Lawrence Erlbaum.

Zeki, E. (1993). *A Vision of the Brain.* Oxford: Blackwell.

CHAPTER 3

Barton, R. A., and R. I. M. Dunbar. (1997). Evolution of the social brain. In A. Whiten and R. W. Byrne (Eds.), *Machiavellian Intelligence 2* (pp. 240–63). Cambridge: Cambridge University Press.

Boysen, S. T. (1996). "More is less": The distribution of rule-governed resource distribution in chimpanzees. In A. E. Russon, K. A. Bard, and S. T. Parker (Eds.), *Reaching into Thought: The Minds of the Great Apes* (pp. 177–89). Cambridge: Cambridge University Press.

Boysen, S. T., and G. G. Berntson. (1995). Responses to quantity: Perceptual versus cognitive mechanisms in chimpanzees (*Pan troglodytes*). *Journal of Comparative Psychology, 21,* 82–86.

Boysen, S. T., and E. J. Capaldi. (1993). *The Development of Numerical Competence: Animal and Human Models.* Hillsdale: Lawrence Erlbaum.

Clark, A. P., and R. W. Wrangham. (1993). Acoustic analysis of wild chimpanzee hoots: Do Kibale Forest chimpanzees have an acoustically distinct food arrival pant hoot? *American Journal of Primatology, 31,* 99–110.

Clutton-Brock, T. H. (1992). *The Evolution of Parental Care.* Princeton: Princeton University Press.

Dantzig, T. (1954). *Number: The Language of Science.* The Free Press: New York.

Davis, H., and R. Perusse. (1988). Numerical competence in animals: Definitional issues, current evidence, and new research agenda. *Behavioral and Brain Sciences, 11,* 561–615.

Dehaene, S. (1997). *The Number Sense.* Oxford: Oxford University Press.

de Waal, F. B. M. (1982). *Chimpanzee Politics: Power and Sex among Apes.* New York: Harper and Row.

———. (1989). *Peacemaking among Primates.* Cambridge: Cambridge University Press.

———. (1996). *Good Natured.* Cambridge: Harvard University Press.

Dugatkin, L. A. (1997). *Cooperation among Animals: An Evolutionary Perspective.* New York: Oxford University Press.

Dunbar, R. (1996). *Grooming, Gossip and the Evolution of Language.* Cambridge: Harvard University Press.

Gallistel, C. R. (1990). *The Organization of Learning.* Cambridge: MIT Press.

Gallistel, C. R., and R. Gellman. (1992). Preverbal and verbal counting and computation. *Cognition, 44,* 43–74.

Gardner, R. A., and B. T. Gardner. (1969). Teaching sign language to a chimpanzee. *Science, 165,* 664–72.

Gardner, R. A., B. T. Gardner, and E. Van Cantfort. (Eds.). (1989). *Teaching Sign Language to Chimpanzees*. Albany: State University of New York Press.

Gellman, R., and C. R. Gallistel. (1986). *The Child's Understanding of Number*. Cambridge: Harvard University Press.

Giraldeau, L. A. (1997). The ecology of information. In J. R. Krebs and N. B. Davies (Eds.), *Behavioural Ecology* (pp. 42–68). Oxford: Blackwell Scientific.

Goodall, J. (1986). *The Chimpanzees of Gombe: Patterns of Behavior*. Cambridge: Harvard University Press.

Grinnell, J., C. Packer, and A. E. Pusey. (1995). Cooperation in male lions: Kinship, reciprocity or mutualism? *Animal Behaviour, 49*, 95–105.

Grinnell, J., and K. McComb. (1996). Maternal grouping as a defense against infanticide by males: Evidence from field playback experiments on African lions. *Behavioral Ecology, 7*, 55–59.

Harcourt, A. H., and F. B. M. de Waal. (1992). *Coalitions and Alliances in Humans and Other Animals*. Oxford: Oxford University Press.

Hauser, M. D. (1997). Tinkering with minds from the past. In M. Daly (Ed.), *Characterizing Human Psychological Adaptations* (pp. 95–131). New York: Wiley.

Hauser, M. D., and S. Carey. (1998). Building a cognitive creature from a set of primitives: Evolutionary and developmental insights. In D. D. Cummins and C. Allen (Eds.), *The Evolution of Mind* (pp. 51–106). Oxford: Oxford University Press.

Hauser, M. D., P. Teixidor, L. Field, and R. Flaherty. (1993). Food-elicited calls in chimpanzees: Effects of food quantity and divisibility? *Animal Behaviour, 45*, 817–19.

Hauser, M. D., and R. W. Wrangham. (1987). Manipulation of food calls in captive chimpanzees: a preliminary report. *Folia primatologica, 48*, 24–35.

Herman, L. M., A. A. Pack, and M. S. Palmer. (1993). Representational and conceptual skills of dolphins. In H. L. Roitblat, L. M. Herman, and P. E. Nachtigall (Eds.), *Language and Communication: Comparative Perspectives* (pp. 403–42). Hillsdale: Lawrence Erlbaum.

Herman, L. M., D. G. Richards, and J. P. Wolz. (1984). Comprehension of sentences by bottlenosed dolphins. *Cognition, 16*, 129–219.

Kahneman, D., A. Treisman, and B. Gibbs. (1992). The reviewing of object files: Object specific integration of information. *Cognitive Psychology, 24*, 175–219.

Karmiloff-Smith, A. (1992). *Beyond Modularity*. Cambridge: MIT Press.

Matsuzawa, T. (1985). Use of numbers by a chimpanzee. *Nature, 315*, 57–59.

———. (1996). Chimpanzee intelligence in nature and in captivity: Isomorphism of symbol use and tool use. In W. C. McGrew, L. F Nishida, and T. Nishida (Eds.), *Great Ape Societies* (pp. 196–209). Cambridge: Cambridge University Press.

McComb, K., C. Packer, and A. Pusey. (1994). Roaring and numerical assessment in contests between groups of female lions, *Panthera leo. Animal Behaviour, 47*, 379–87.

McComb, K., A. Pusey, C. Packer, and J. Grinnell. (1993). Female lions can identify potentially infanticidal males from their roars. *Proceedings of the Royal Society, London, 252*, 59–64.

Meck, W. H., and R. M. Church. (1983). A mode control model of counting and timing processes. *Journal of Experimental Psychology: Animal Behavior Processes, 2*, 320–34.

Packer, C., and A. E. Pusey. (1983). Adaptations of female lions to infanticide by incoming males. *American Naturalist, 121*, 716–28.

Packer, C., and L. Ruttan. (1988). The evolution of cooperative hunting. *American Naturalist*, *132*, 159–98.

Pepperberg, I. M. (1987). Acquisition of the same/different concept by an African Grey parrot (*Psittacus erithacus*): Learning with respect to categories of color, shape, and material. *Journal of Experimental Analysis of Behavior*, *50*, 553–64.

———. (1991). A communicative approach to animal cognition: A study of conceptual abilities of an African grey parrot. In C. A. Ristau (Ed.), *Cognitive Ethology: The Minds of Other Animals* (pp. 153–86). Hillsdale: Lawrence Erlbaum.

———. (1994). Numerical competence in an African gray parrot (*Psittacus erithacus*). *Journal of Comparative Psychology*, *108*, 36–44.

Pfungst, O. (1911). *Clever Hans*. New York: Henry Holt.

Premack, D. (1986). *Gavagai! or the Future History of the Animal Language Controversy*. Cambridge: MIT Press.

Premack, D., and A. Premack. (1994). Levels of causal understanding in chimpanzees and children. *Cognition*, *50*. 347–62.

Premack, D., and G. Woodruff. (1978). Does the chimpanzee have a theory of mind? *Behavioral and Brain Sciences*, *4*, 515–26.

Rumbaugh, D. M., and D. A. Washburn. (1993). Counting by chimpanzees and ordinality judgements by macaques in video-formatted tasks. In S. T. Boysen and E. J. Capaldi (Eds.), *The Development of Numerical Competence: Animal and Human Models* (pp. 87–108). Hillsdale: Lawrence Erlbaum.

Savage-Rumbaugh, E. S. (1986). *Ape Language: From Conditioned Response to Symbol*. New York: Columbia University Press.

Schusterman, R. J., R. Gisiner, B. K. Grimm, and E. B. Hanggi. (1993). Behavior control by exclusion and attempts at establishing semanticity in marine mammals using match-to-sample paradigms. In H. L. Roitblat, L. M. Herman, and P. E. Nachtigall (Eds.), *Language and Communication: Comparative Perspectives* (pp. 249–74). Hillsdale: Lawrence Erlbaum.

Sebeok, T. A., and R. Rosenthal. (1981). The Clever Hans phenomenon: Communication with horses, whales, apes, and people. *Annals of the New York Academy of Sciences*, *364*, 1–311.

Spelke, E. S., A. T. Phillips, and A. L. Woodward. (1995). Infants' knowledge of object motion and human action. In A. J. Premack, D. Premack, and D. Sperber (Eds.), *Causal Cognition: A Multidisciplinary Debate* (pp. 44–77). Oxford: Clarendon Press.

Starkey, P., and R. Cooper. (1980). Perception of numbers by human infants. *Science*, *210*, 1033–1035.

Stephens, D. W., and J. R. Krebs. (1986). *Foraging Theory*. Princeton: Princeton University Press.

Terrace, H. S. (1979). *Nim*. New York: Knopf.

Trick, L., and Z. Pylyshyn. (1994). Why are small and large numbers enumerated differently? A limited capacity preattentive stage in vision. *Psychological Review*, *101*, 80–102.

Uller, C. (1997). *Origins of Numerical Concepts: A Comparative Study of Human Infants and Nonhuman Primates*. Unpublished Ph.D., MIT, Cambridge.

Washburn, D. A., and D. M. Rumbaugh. (1991). Ordinal judgements of numerical symbols by macaques (*Macaca mulatta*). *Psychological Science*, *2*, 190–93.

Wrangham, R. W. (1977). Feeding behaviour of chimpanzees in Gombe National Park, Tanzania. In T. H. Clutton-Brock (Ed.), *Primate Ecology: Studies of Feeding and Ranging Behaviour in Lemurs, Monkeys and Apes* (pp. 504–38). London: Academic Press.

Wrangham, R. W., and D. Peterson. (1996). *Demonic Males.* New York: Houghton Mifflin.

Wynn, K. (1990). Children's understanding of counting. *Cognition, 36,* 155–93.

———. (1992). Addition and subtraction by human infants. *Nature, 358,* 749–50.

———. (1995). Origins of numerical knowledge. *Mathematical Cognition, 1,* 35–60.

CHAPTER 4

Able, K. P. (1991). Common themes and variations in animal orientation systems. *American Zoologist, 31,* 157–67.

———. (1996). The debate over olfactory navigation by homing pigeons. *Journal of Experimental Biology, 199,* 121–24.

Balda, R. P., and A. C. Kamil. (1989). A comparative study of cache recovery by three corvid species. *Animal Behaviour, 38.*

Basil, J. A., A. C. Kamil, R. P. Balda, and K.V. Fite. (1996). Differences in hippocampal volume among food storing corvids. *Brain, Behavior and Evolution, 47,* 156–64.

Bennett, A. T. D. (1996). Do animals have cognitive maps? *Journal of Experimental Biology, 199,* 219–24.

Biegler, R., and R. G. M. Morris. (1996). Landmark stability: Studies exploring whether the perceived stability of the environment influences spatial representation. *Journal of Experimental Biology, 199,* 187–93.

Boesch, C., and H. Boesch. (1984). Mental map in wild chimpanzees: An analysis of hammer transports for nut cracking. *Primates, 25,* 160–70.

Chappell, J., and T. Guilford. (1995). Homing pigeons primarily use the sun compass rather than fixed directional visual cues in an open-field arena food-searching task. *Proceedings of the Royal Society, London, B, 260,* 59–63.

Cheng, K. (1986). A purely geometric module in the rat's spatial representation. *Cognition, 23,* 149–78.

Clayton, N. S. (1995). The neuroethological development of food-storing memory: A case of use it, or lose it! *Behavioural Brain Research, 70,* 95–102.

Clayton, N. S., and J. R. Krebs. (1994). Hippocampal growth and attrition in birds affected by experience. *Proceedings of the National Academy of Sciences, 91,* 7410–7414.

Collett, T. S. (1996). Insect navigation en route to the goal: Multiple strategies for the use of landmarks. *Journal of Experimental Biology, 199,* 227–35.

Cosmides, L., and J. Tooby. (1994). Origins of domain specificity: The evolution of functional organization. In L. A. Hirschfeld and S. A. Gelman (Eds.), *Mapping the Mind: Domain-Specificity in Cognition and Culture* (pp. 85–116). Cambridge: Cambridge University Press.

DeLoache, J. S., and A. L. Brown. (1983). Very young children's memory for the location of objects in a large-scale environment. *Child Development, 54,* 888–97.

Dittman, A. H., and T. P. Quinn. (1996). Homing in pacific salmon: Mechanisms and ecological basis. *Journal of Experimental Biology, 199,* 83–91.

Duff, S. J., L. A. Brownlie, and D. F. Sherry. (1999). Sun compass and landmark orientation by black-capped chickadees (*Parus atricapillus*). *Journal of Experimental Psychology: Animal Behavior Processes, 24,* 243–53.

Dyer, F. C. (1996). Spatial memory and navigation by honeybees on the scale of the foraging range. *Journal of Experimental Biology*, 199, 147–54.

Emlen, S. T. (1975). Migration: Orientation and navigation. In D. S. Farrier and J. R. King (Eds.), *Avian Biology* (pp. 129–219). New York: Academic Press.

Gallistel, C. R. (1990). *The Organization of Learning*. Cambridge: MIT Press.

———. (1998). Insect navigation: brains as symbol processors. In S. Sternberg and D. Scarborough (Eds.), *Conceptual and Methodological Foundations. Volume 4 of An Invitation to Cognitive Science* (pp. 1–52). Cambridge: MIT Press.

Gallistel, C. R., and A. E. Cramer. (1996). Computations on metric maps in mammals: Getting oriented and choosing a multi-destination route. *Journal of Experimental Biology*, 199, 211–17.

Gaulin, S. J. C., and R. W. FitzGerald. (1986). Sex differences in spatial ability: An evolutionary hypothesis and test. *American Naturalist*, 127, 74–88.

———. (1989). Sexual selection for spatial-learning ability. *Animal Behaviour*, 37, 322–31.

Gould, J., and C. Gould. (1988). *The Honey Bee*. New York: Scientific American.

Gould, J., and C. Gould. (1995). *The Animal Mind*. New York: Scientific American.

Gwinner, E. (1996). Circadian and circannual programmes in avian migration. *Journal of Experimental Biology*, 199, 39–48.

Healy, S. D. (1996). Ecological specialization in the avian brain. In C. F. Moss and S. J. Shettleworth (Eds.), *Neuroethological Studies of Cognitive and Perceptual Processes*, (pp. 84–112). Boulder: Westview.

Hermer, L., and E. S. Spelke. (1994). A geometric process for spatial reorientation in young children. *Nature*, 370, 57–59.

———. (1996). Modularity and development: The case of spatial reorientation. *Cognition*, 61, 195–232.

Jacobs, L. F., and W. D. Spencer. (1994). Natural space-use patterns and hippocampal size in kangaroo rats. *Brain, Behavior and Evolution*, 44, 125–32.

Joubert, A., and J. Vauclair. (1985). Reaction to novel objects in a troop of guinea baboons: Approach and manipulation. *Behaviour*, 73, 92–104.

Judd, S. P. D., and T. S. Collett. (1998). Multiple stored views and landmark guidance in ants. *Nature*, 392, 710–14.

Kamil, A. C., and J. E. Jones. (1997). The seed-storing corvid Clark's nutcracker learns geometric relationships among landmarks. *Nature*, 390, 276–79.

Krebs, J. R., D. F. Sherry, S. D. Healy, V. H. Perry, and A. L. Vaccarino. (1989). Hippocampal specialization of food-storing in birds. *Proceedings of the National Academy of Sciences*, 86, 1388–1392.

Landau, B., E. Spelke, and H. Gleitman. (1984). Spatial knowledge in a young blind child. *Cognition*, 16, 225–60.

Margules, J., and C. R. Gallistel. (1988). Heading in the rat: Determination by environmental shape. *Animal Learning and Behavior*, 16, 404–10.

Menzel, E. (1973). Chimpanzee spatial memory organization. *Science*, 182, 943–45.

Menzel, E. W., and C. R. Menzel. (1979). Cognitive, developmental and social aspects of responsiveness to novel objects in a family group of marmosets (*Saguinus fuscicollis*). *Behaviour*, 70, 251–79.

Menzel, R., K. Geiger, L. Chittka, J. Joerges, J. Junze, and U. Muller. (1996). The knowledge base of bee navigation. *Journal of Experimental Biology*, 199, 141–46.

O'Keefe, J., and L. Nadel. (1978). *The Hippocampus as a Cognitive Map*. Oxford: Clarendon Press.

Pinker, S. (1997). *How the Mind Works*. New York: Norton.

Poucet, B. (1993). Spatial cognitive maps in animals: New hypotheses on their structure and neural mechanisms. *Psychological Review, 100*, 163–82.

Premack, D., and A. Premack. (1983). *The Mind of an Ape*. New York: Norton.

Santchi, F. (1913). Comment s'orient les fourmis. *Revue Suisse de Zoologie, 21*, 347–426.

Savage-Rumbaugh, E. S. (1986). *Ape Language: From Conditioned Response to Symbol*. New York: Columbia University Press.

Savage-Rumbaugh, E. S., and R. Lewin. (1996). *Kanzi*. New York: Wiley.

Savage-Rumbaugh, E. S., K. McDonald, R. A. Sevcik, W. D. Hopkins, and E. Rubert. (1986). Spontaneous symbol acquisition and communicative use by pygmy chimpanzees (*Pan paniscus*). *Journal of Experimental Psychology: General, 115*, 211–35.

Sherry, D. F. (1997). Cross-species comparisons. In M. Daly (Ed.), *Characterizing Human Psychological Adaptations* (pp. 181–94). New York: Wiley.

Sherry, D. F., M. R. L. Forbes, M. Khurgel, and G. O. Ivy. (1993). Females have a larger hippocampus than males in the brood-parasitic brownheaded cowbird. *Proceedings of the National Academy of Sciences, 90*, 7839–7843.

Sherry, D. F., and E. Hampson. (1997). Evolution and the hormonal control of sexually dimorphic spatial abilities in humans. *Trends in Cognitive Science, 1*, 50–56.

Sherry, D. F., L. F. Jacobs, and S. J. C. Gaulin. (1992). Spatial memory and adaptive specializations of the hippocampus. *Trends in Neurosciences, 15*, 298–303.

Sherry, D. F., and A. L. Vaccarino. (1989). Hippocampus and memory for food caches in black-capped chickadees. *Behavioral Neuroscience, 103*, 308–13.

Sherry, D. F., A. L. Vaccarino, K. Buckenham, and R. S. Herz. (1989). The hippocampal complex of food-storing birds. *Brain, Behavior and Evolution, 34*, 308–17.

Shettleworth, S. (1998). *Cognition, Evolution and Behavior*. New York: Oxford University Press.

Steiger, B., and S. H. Steiger. (1992). *Strange Powers of Pets*. New York: D. I. Fine.

Thinus-Blanc, C. (1995). Spatial information processing in animals. In H. L. Roitblat and J. A. Meyer (Eds.), *Comparative Approaches to Cognitive Science* (pp. 241–69). Cambridge: MIT Press.

Tolman, E. C. (1955). Cognitive maps in rats and men. *Psychological Review, 55*, 189–208.

Walker, M. M., C. E. Diebel, C. V. Haugh, P. M. Pankhurst, J.C. Montgomery, and C. R. Greer. (1997). Structure and function of the vertebrate magnetic sense. *Nature, 390*, 371–76.

Wallraff, H. G. (1996). Seven theses on pigeon homing deduced from empirical findings. *Journal of Experimental Biology, 199*, 105–11.

Wehner, R. (1997). Sensory systems and behaviour. In J. R. Krebs and N. B. Davies (Eds.), *Behavioural Ecology* (pp. 19–41). Oxford: Blackwell Scientific.

Wehner, R., and R. Menzel. (1990). Do insects have cognitive maps? *Annual Review of Neuroscience, 13*, 403–14.

Wehner, R., B. Michel, and P. Antonsen. (1996). Visual navigation in insects: Coupling of egocentric and geocentric information. *Journal of Experimental Biology, 199*, 129–40.

Wiltschko, W., and R. P. Balda. (1989). Sun compass orientation in seed-catching scrub jays (*Aphelocoma coerulescens*). *Journal of Comparative Physiology, A, 164*, 717–21.

Wiltschko, W., and R. Wiltschko. (1996). Magnetic orientation in birds. *Journal of Experimental Biology*, *199*, 29–38.

CHAPTER 5

Anderson, J. R. (1984). Monkeys with mirrors: Some questions for primate psychology. *International Journal of Primatology*, *5*, 81–98.

Bateson, P. P. G. (1966). The characteristics and context of imprinting. *Biological Reviews*, 41, 177–220.

Beecher, M. D. (1982). Signature systems and kin recognition. *American Zoologist*, *22*, 477–90.

Bermudez, J. L., A. Marcel, and N. Eilan (Eds.). (1995). *The Body and the Self*. Cambridge: MIT Press.

Bouchard, T. J., Jr. (1997). Whenever the twain shall meet. *The Sciences*, *10*, 52–57.

Bourke, A. F. G. (1997). Sociality and kin selection in insects. In J. R. Krebs and N. B. Davies (Eds.), *Behavioural Ecology* (pp. 203–27). Oxford: Blackwell Scientific.

Brown, W. L., A. A. McDowell, and E. M. Robinson. (1965). Discrimination learning of mirrored cues by rhesus monkeys. *Journal of Genetic Psychology*, *106*, 123–28.

Burghardt, G. M. (1991). Cognitive ethology and critical anthropomorphism: A snake with two heads and hognose snakes that play dead. In C. A. Ristau (Ed.), *Cognitive Ethology: The Minds of Other Animals* (pp. 53–91). Hillsdale: Lawrence Erlbaum.

Damasio, A. (1994). *Descartes' Error: Emotion, Reason and the Human Brain*. New York: Putnam.

Damasio, A. R., H. Damasio, and, D. Tranel. (1990a). Impairments of visual recognition as clues, to the processes of memory. In G. M. Edelman, W. E. Gall, and W. M. Cowan (Eds.), *Signal and Sense: Local and Global Order in Perceptual Maps* (pp. 451–73). New York: Wiley-Liss.

Damasio, A. R., D. Tranel, and H. Damasio. (1990b). Face agnosia and the neural substrate of memory. *Annual Review of Neuroscience*, *13*, 89–109.

Darwin, C. (1871). *The Descent of Man and Selection in Relation to Sex*. London: John Murray.

———. (1872). *The Expression of the Emotions in Man and Animals*. London: John Murray.

Dawkins, R. (1976). *The Selfish Gene*. Oxford: Oxford University Press.

Dooling, R. J., S. D. Brown, K. Manabe, and E. F. Powell. (1996). The perceptual foundations of vocal learning in budgerigars. In C. F. Moss and S. J. Shettleworth (Eds.), *Neuroethological Studies of Cognitive and Perceptual Processes* (pp. 113–37). Boulder: Westview.

Ellis, H. D., and A. W. Young. (1990). Accounting for delusional misidentification. *British Journal of Psychiatry*, *157*, 239–48.

Epstein, R., R. P. Lanza, and B. F. Skinner. (1981) "Self -awareness" in the pigeon. *Science*, *212*, 645–96.

Fouts, R., and S. T. Mills. (1997). *Next of Kin*. New York: Morrow.

Gallup, G. G., Jr. (1970). Chimpanzees: Self-recognition. *Science*, *167*, 86–87.

———. (1987). Self-awareness. In J. R. Erwin and G. Mitchell (Eds.), *Comparative Primate Biology, Volume 2B: Behavior, Cognition, and Motivation* (pp. 3–16). New York: Alan Liss.

———. (1991). Toward a comparative psychology of self-awareness: Species limitations and cognitive consequences. In G. R. Goethals and J. Strauss (Eds.), *The Self: An Interdisciplinary Approach* (pp. 121–35). New York: Springer-Verlag.

Gallup, G. G., Jr., and D. Povinelli. (1993). Mirror, mirror on the wall which is the most heuristic theory of them all? A response to Mitchell. *New Ideas in Psychology, 11,* 327–35.

Grosberg, R. K. (1988). The evolution of allo-recognition specificity in clonal vertebrates. *Quarterly Review of Biology, 63,* 377–412.

Hauser, M. D., J. Kralik, C. Botto, M. Garrett, and J. Oser. (1995). Self-recognition in primates: Phylogeny and the salience of species-typical traits. *Proceedings of the National Academy of Sciences, 92,* 10811–10814.

Hirstein, W., and V. S. Ramachandran. (1997). Capgras syndrome: A novel probe for understanding the neural representation of the identity and familiarity of persons. *Proceedings of the Royal Society, London, 264,* 427–44.

Horn, G. (1991). Cerebral function and behaviour investigated through a study of filial imprinting. In P. Bateson (Ed.), *The Development and Integration of Behaviour* (pp. 121–48). Cambridge: Cambridge University Press.

Johnson, M. H., and J. Morton. (1991). *Biology and Cognitive Development.* Cambridge: Basil Blackwell.

Lorenz, K. (1937). The companion in the bird's world. *Auk, 54,* 245–73.

Margoliash, D. (1983). Acoustic parameters underlying the responses of song-specific neurons in the white-crowned sparrow. *Journal of Neuroscience, 3,* 1039–1057.

———. (1987). Preference for autogenous song by auditory neurons in a song system nucleus of the white-crowned sparrow. *Journal of Neuroscience, 6,* 1643–1661.

Marler, P. (1970). A comparative approach to vocal learning: Song development in white-crowned sparrows. *Journal of Comparative Physiological Psychology Monographs, 71,* 1–25.

Marler, P., and M. Tamura. (1962). Culturally transmitted patterns of vocal behavior in sparrows. *Science, 146,* 1483–1486.

Maynard Smith, J. (1964). Group selection and kin selection. *Nature, 201,* 1145–1147.

McGregor, P. K. (Ed.). (1992). *Playback and Studies of Animal Communication.* New York: Plenum.

Menzel, E. W., E. S. Savage-Rumbaugh, and J. Lawson. (1985). Chimpanzee (*Pan troglodytes*) spatial problem solving with the use of mirrors and televised equivalents of mirrors. *Journal of Comparative Psychology, 99,* 211–217.

Mitchell, R. W. (1993). Mental models of mirror-self-recognition: Two theories. *New Ideas in Psychology, 11,* 295–325.

Moody, M. F. (1975). Perception of total reflection in *Barbus. Behavioral Biology, 15,* 239–43.

Pepperberg, I. M., S. E. Garcia, E. C. Jackson, and S. Marconi. (1995). Mirror use by African Grey Parrots (*Psittacus erithacus*). *Journal of Comparative Psychology, 109,* 182–95.

Povinelli, D. J. (1987). Monkeys, apes, mirrors, and minds: The evolution of self-awareness in primates. *Human Evolution, 2,* 493–507.

———. (1989). Failure to find self-recognition in Asian elephants (*Elephas maximus*) in contrast to their use of mirror cues to discover hidden food. *Journal of Comparative Psychology, 103,* 122–31.

———. (1993). Reconstructing the evolution of mind. *American Psychologist, 48,* 493–509

Povinelli, D. J., A. B. Rulf, K. R. Landau, and D. T. Bierschwale. (1993). Self-recognition in chimpanzees (*Pan troglodytes*): Distribution, ontogeny, and patterns of emergence. *Journal of Comparative Psychology, 107,* 347–72.

Ramachandran, V. S., and S. Blakeslee. (1998). *Phantoms in the Brain.* New York: William Morrow.

Sherman, P. W., H. K. Reeve, and D. W. Pfennig. (1997). Recognition systems. In J. R. Krebs and N. B. Davies (Eds.), *Behavioural Ecology* (pp. 69–96). Oxford: Blackwell Scientific.

Swartz, K. B., and S. Evans. (1991). Not all chimpanzees (*Pan troglodytes*) show self-recognition. *Primates, 32,* 483–96.

———. (1997). Anthropomorphism, anecdotes, and mirrors. In R. W. Mitchell, N. S. Thompson, and H. L. Miles (Eds.), *Anthropomorphism, Anecdotes, and Animals* (pp. 296–312). Albany: SUNY.

Tovee, M. J., and E. M. Cohen-Tovee. (1993). The neural substrate of face processing models: A review. *Cognitive Neuropsychology, 10,* 505–28.

Tranel, D., and A. R. Damasio. (1985). Knowledge without awareness: An autonomic index of recognition of prosapagnosics. *Science, 228,* 1453–1454.

———. (1993). The covert learning of affective valence does not require structures in hippocampal system of amygdala. *Journal of Cognitive Neuroscience, 5,* 79–88.

Tranel, D., A. R. Damasio, and H. Damasio. (1988). Intact recognition of facial expression, gender, and age in patients with impaired recognition of face identity. *Neurology, 38,* 690–96.

Young, A. W. (1992). Face recognition impairments. *Philosophical Transactions of the Royal Society of London, 335,* 47–54.

Chapter 6

Alvarez-Buylla, A., J. R. Kirn, and F. Nottenbohm. (1990). Birth of projection neurons in adult avian brain may be related to perceptual or motor learning. *Science, 249,* 1444–1446.

Andersson, M. (1994). *Sexual Selection.* Princeton: Princeton University Press.

Baker, A. J., and P. F. Jenkins. (1987). Founder effect and cultural evolution of songs in an isolated population of chaffinches, *Fringilla coelebs,* in the Chatham Islands. *Animal Behaviour, 35,* 1179–1803.

Boesch, C. (1991). Teaching among wild chimpanzees. *Animal Behaviour, 41,* 530–32.

———. (1996). Three approaches for assessing chimpanzee culture. In A. E. Russon, K. A. Bard, and S. T. Parker (Eds.), *Reaching into Thought: The Minds of the Great Apes* (pp. 404–29). Cambridge: Cambridge University Press.

Boyd, R, and P. J. Richerson. (1985). *Culture and the Evolutionary Process.* Chicago: University of Chicago Press.

Brenowitz, E. A., and D. E. Kroodsma. (1996). The neuroethology of birdsong. In D. E. Kroodsma and E. H. Miller (Eds.), *Ecology and Evolution of Acoustic Communication in Birds* (pp. 285–304). Ithaca: Cornell University Press.

Byme, R. (1995). *The Thinking Ape.* Oxford: Oxford University Press.

Candland, D. K. (1995). *Feral Children and Clever Animals.* Oxford: Oxford University Press.

Caro, T. M. (1980a). Effects of the mother, object play and adult experience on predation in cats. *Behavioral and Neural Biology*, 29, 29–51.

———. (1980b). Predatory behaviour in domestic cat mothers. *Behaviour*, 74, 128–47.

———. (1994). *Cheetahs of the Serengeti Plains: Grouping in an Asocial Species*. Chicago: The University of Chicago Press.

Caro, T. M., and M.D. Hauser. (1992). Is there teaching in nonhuman animals? *Quarterly Review of Biology*, 67, 151–74.

Catchpole, C. K., and P. Slater. (1995). *Bird Song: Biological Themes and Variations*. Cambridge: Cambridge University Press.

Cheney, D. L., and R. M. Seyfarth. (1990). *How Monkeys See the World: Inside the Mind of Another Species*. Chicago: University of Chicago Press.

Clutton-Brock, T. H., and K. McComb. (1993). Experimental tests of copying and mate choice in fallow deer (*Dama dama*). *Behavioral Ecology*, 4, 191–93.

Cosmides, L., and J. Tooby. (1992). Cognitive adaptations for social exchange. In J. Barkow, L. Cosmides, and J. Tooby (Eds.), *The Adapted Mind* (pp. 163–228). New York: Oxford University Press.

———. (1994). Origins of domain specificity: The evolution of functional organization. In L. A. Hirschfeld and S. A. Gelman (Eds.), *Mapping the Mind: Domain-Specificity in Cognition and Culture* (pp. 85–116). Cambridge: Cambridge University Press.

Cristol, D. A., P. V. Switzer, K. L. Johnson, and L. S. Walke. (1998). Crows do not use automobiles as nutcrackers: Putting an anecdote to the test. *Auk*.

Custance, D. M., A. Whiten, and K. A. Bard. (1995). Can young chimpanzees (*Pan troglodytes*) imitate arbitrary actions? Hayes & Hayes (1952) revisited. *132*.

Daly, M. (Ed.) (1997). *Characterizing Human Psychological Adaptations*. New York: Wiley.

Darwin, C. (1871). *The Descent of Man and Selection in Relation to Sex*. London: John Murray.

Dugatkin, L. (1992). Sexual selection and imitation: Females copy the mate choice of others. *American Naturalist*, 139, 1384–1389.

———. (1997). Copying and mate choice. In C. M. Heyes and B. G. Galef, Jr. (Eds.), *Social Learning in Animals: The Roots of Culture* (pp. 85–107). San Diego: Academic Press.

Fiorito, G., and P. Scotto. (1992). Observational learning in *Octopus vulgaris*. *Science*, 256, 545–47.

Fisher, J., and R. A. Hinde. (1949). The opening of milk bottles by birds. *British Birds*, 42, 347–57.

Fitch, W. T. (1997). Vocal tract length and formant frequency dispersion correlate with body size in rhesus macaques. *Journal of the Acoustical Society of America*, 102, 1213–1222.

Fitch, W. T., and M. D. Hauser. (1995). Vocal production in nonhuman primates: Acoustics, physiology and functional constraints on honest advertisement. *American Journal of Primatology*, 37, 191–219.

Fragaszy, D. M., and E. Visalberghi. (1996). Social learning in monkeys: Primate "primacy" reconsidered. In C. M. Heyes and B. G. Galef, Jr. (Eds.), *Social Learning in Animals: The Roots of Culture* (pp. 65–84). San Diego: Academic Press.

Galef, B. G., Jr. (1988). Imitation in animals: History, definitions, and interpretation of data from the psychological laboratory. In T. Zentall and B. G. Galef (Eds.), *Social

Learning: Psychological and Biological Perspectives (pp. 3–28). Hillsdale: Lawrence Erlbaum.

———. (1996). Social enhancement of food preferences in Norway rats: A brief review. In C. M. Heyes and B. G. Galef, Jr. (Eds.), *Social Learning in Animals: The Roots of Culture* (pp. 49–64). San Diego: Academic Press.

Gibson, R. M., J. W. Bradbury, and S. L. Vehrencamp. (1991). Mate choice in lekking sage grouse: The roles of vocal display, female site fidelity and copying. *Behavioral Ecology, 2,* 165–80.

Hauser, M. D. (1996). *The Evolution of Communication.* Cambridge: MIT Press.

Hayes, K. J., and C. Hayes. (1952). Imitation in the home-raised chimpanzee. *Journal of Comparative and Physiological Psychology, 45,* 450–59.

Heyes, C. M. (1996). Genuine imitation? In C. M. Heyes and B. G. Galef, Jr. (Eds.), *Social Learning in Animals: The Roots of Culture* (pp. 371–90). San Diego: Academic Press.

Heyes, C. M., and G. R. Dawson. (1990). A demonstration of observational learning using a bidirectional control. *Quarterly Journal of Experimental Psychology, 42B,* 59–71.

Heyes, C. M., and B. G. Galef, Jr. (Eds.). (1996). *Social Learning in Animals: The Roots of Culture.* San Diego: Academic Press.

Heyes, C. M., E. Jaldow, T. Nokes, and G. R. Dawson. (1994). Imitation in rats: Conditions of occurrence in a bi-directional control procedure. *Learning and Motivation, 25,* 276–87.

Houde, A. E., and J. A. Endler. (1990). Correlated evolution of female mating preferences and male color patterns in the guppy, *Poecilia reticulata. Science, 248,* 1405–1408.

Ingmanson, E. J. (1996). Tool-using behavior in wild *Pan paniscus*: Social and ecological considerations. In A. E. Russon, K. A. Bard, and S. T. Parker (Eds.), *Reaching into Thought: The Minds of the Great Apes* (pp. 190–210). Cambridge: Cambridge University Press.

Jung, C. G. (1953). *Psychology and Alchemy.* Boston: Little Brown.

Kirkpatrick, M., and M. J. Ryan. (1991). The paradox of the lek and the evolution of mating preferences. *Nature, 350,* 33–38.

Konishi, M. (1985). Birdsong: From behavior to neuron. *Annual Review of Neuroscience, 8,* 125–70.

Kroodsma, D. E., and E. H. Miller (Eds.). (1996). *Ecology and Evolution of Acoustic Communication in Birds.* Ithaca: Comstock.

Laland, K. N., P. J. Richerson, and R. Boyd. (1996). Developing a theory of animal social learning. In C. M. Heyes and B. G. Galef, Jr. (Eds.), *Social Learning in Animals: The Roots of Culture* (pp. 129–54). San Diego: Academic Press.

Lieberman, P. (1984). *The Biology and Evolution of Language.* Cambridge: Harvard University Press.

Locke, J. (1993). *The Path to Spoken Language.* Cambridge: Harvard University Press.

Marler, P. (1970). A comparative approach to vocal learning: Song development in white-crowned sparrows. *Journal of Comparative Physiological Psychology Monogaraphs, 71,* 1–25.

———. (1989). Learning by instinct: Birdsong. *American Speech-Language Association, 89,* 75–79.

———. (1997). Three models of song learning: Evidence from behavior. *Journal of Neurobiology*, 33, 1–16.

Matsuzawa, T. (in press). Cultural and social contexts of tool use in chimpanzees. In M. D. Hauser and M. Konishi (Eds.), *The Design of Animal Communication* Cambridge: MIT Press.

Meltzoff, A. N. (1988). The human infant as *Homo imitans*. In T. R. Zentall and B. G. Galef, Jr. (Eds.), *Social Learning: Psychological and Biological Perspectives* (pp. 319–41). Hillsdale: Lawrence Erlbaum.

———. (1996). The human infant as imitative generalist: A 20-year progress report on infant imitation with implications for comparative psychology. In C. M. Heyes and B. G. Galef, Jr. (Eds.), *Social Learning in Animals: The Roots of Culture* San Diego: Academic Press.

Meltzoff, A. N., and M. K. Moore. (1995). Infants' understanding of people and things: From body imitation to folk psychology. In J. L. Bermudez, A. Marcel, and N. Eilan (Eds.), *The Body and the Self* (pp. 43–70). Cambridge: MIT Press.

Møller, A. P. (1994). *Sexual Selection and the Barn Swallow*. Oxford: Oxford University Press.

Moore, B. R. (1992). Avian movement imitation and a new form of mimicry: Tracing the evolution of a complex form of learning. *Behaviour*, 122, 231–63.

———. (1996). The evolution of imitative learning. In C. M. Heyes and B. G. Galef, Jr. (Eds.), *Social Learning in Animals: The Roots of Culture* (pp. 245–66). San Diego: Academic Press.

Nishida, T. (1987). Local traditions and cultural tradition. In B. B. Smuts, D. L. Cheney, R. M. Seyfarth, R. W. Wrangham, and T. T. Struhsaker (Eds.), *Primate Societies* (pp. 462–74). Chicago: University of Chicago Press.

Nottebohm, F., S. Kasparian, and C. Pandazis (1981). Brain space for a learned task. *Brain Research*, 213, 99–109.

Parker, S. T. (1996). Apprenticeship in tool-mediated extractive foraging: The origins of imitation, teaching and self-awareness in great apes. In A. E. Russon, K. A. Bard, and S. T. Parker (Eds.), *Reaching into Thought: The Minds of the Great Apes* (pp. 348–70). Cambridge: Cambridge University Press.

Parker, S. T., and A. E. Russon. (1996). On the wild side of culture and cognition in the great apes. In A. E. Russon, K. A. Bard, and S. T. Parker (Eds.), *Reaching into Thought: The Minds of the Great Apes* (pp. 430–50). Cambridge: Cambridge University Press.

Payne, R. B., L. L. Payne, and S. M. Doehlert. (1988). Biological and cultural success of song memes in indigo buntings. *Ecology*, 69, 104–17.

Pinker, S. (1994). *The Language Instinct*. New York: William Morrow.

———. (1997). *How the Mind Works*. New York: Norton.

Ramachandran, V. S., and S. Blakeslee. (1998). *Phantoms in the Brain*. New York: William Morrow.

Rawls, K., P. Fiorelli, and S. Gish. (1985). Vocalizations and vocal mimicry in captive Harbor seals, *Phoca vitulina*. *Canadian Journal of Zoology*, 63, 1050–1056.

Russon, A. E. (1997). Exploiting the expertise of others. In A. Whiten and R. W. Byrne (Eds.), *Machiavellian Intelligence 2* (pp. 174–206). Cambridge: Cambridge University Press.

Russon, A. E., and B. M. F. Galdikas. (1993). Imitation in free-ranging rehabilitant orangutans (*Pongo pygmaeus*). *Journal of Comparative Psychology*, 107, 147–61.

———. (1995). Constraints on Great Apes' imitation: Model and action selectivity in rehabilitant orangutan (*Pongo pygmaeus*) imitation. *Journal of Comparative Psychology*, 109, 5–17.

Sherry, D. F., and B. G. Galef, Jr. (1984). Cultural transmission without imitation: Milk bottle opening by birds. *Animal Behaviour*, 32, 937–38.

———. (1990). Social learning without imitation: More about milk bottle opening by birds. *Animal Behaviour*, 40, 987–89.

Snowdon, C. T., and M. Hausberger (Eds.). (1997). *Social Influences on Vocal Development*. Cambridge: Cambridge University Press.

Tayler, C. K., and G. S. Saayman. (1973). Imitative behaviour by Indian Ocean bottlenose dolphins (*Tursiops aduncus*) in captivity. *Behaviour*, 44, 286–98.

Terkel, J. (1996). Cultural transmission of feeding behavior in the black rat (*Rattus rattus*). In C. M. Heyes and B. G. Galef, Jr. (Eds.), *Social Learning in Animals: The Roots of Culture* (pp. 17–48). San Diego: Academic Press.

Tomasello, M. (1996). Do apes ape? In C. M. Heyes and B. G. Galef, Jr. (Eds.), *Social Learning in Animals: The Roots of Culture* (pp. 319–46). San Diego: Academic Press.

Tomasello, M., and J. Call. (1997). *Primate Cognition*. Oxford: Oxford University Press.

Tomasello, M., A. Kruger, and H. Ratner. (1993). Cultural learning. *Behavioral and Brain Sciences*, 16, 495–552.

Visalberghi, E., and D. Fragaszy. (1991). Do monkeys ape? In S. T. Parker and K. R. Gibson (Eds.), *"Language" and Intelligence in Monkeys and Apes* (pp. 247–73). Cambridge: Cambridge University Press.

Warner, R. R. (1988). Traditionality of mating-site preferences in coral reef fish. *Nature*, 335, 719–21.

West, M. J., and A. P. King. (1996). Social learning: Synergy and songbirds. In C. M. Heyes and B. G. Galef, Jr. (Eds.), *Social Learning in Animals: The Roots of Culture* (pp. 155–78). San Diego: Academic Press.

Whiten, A., and D. Custance. (1996). Studies of imitation in chimpanzees and children. In C. M. Heyes and B. G. Galef, Jr. (Eds.), *Social Learning in Animals: The Roots of Culture* (pp. 291–318). San Diego: Academic Press.

CHAPTER 7

Baron-Cohen, S. (1995). *Mindblindness*. Cambridge: MIT Press.

Baron-Cohen, S., H. Tager-Flusberg, and D. Cohen. (Eds.). (1993). *Understanding Other Minds: Perspectives from Autism*. Oxford: Oxford University Press.

Butterworth, G., and E. Cochran. (1991). What minds have in common is space: Spatial mechanisms serving joint visual attention in infancy. *British Journal of Developmental Psychology*, 9, 55–72.

Byrne, R., and A. Whiten. (1990). Tactical deception in primates: The 1990 database. *Primate Report*, 27, 1–101.

Caldwell, R. L. (1986). The deceptive use of reputation by stomatopods. In R. W. Mitchell and N. S. Thompson (Eds.), *Deception: Perspectives on Human and Nonhuman Deceit*. Albany: SUNY Press.

Caldwell, R. L., and H. Dingle. (1975). Ecology and evolution of agonistic behavior in the stomatopods. *Naturwissenschaften*, 62, 214–22.

Carruthers, P., and P. K. Smith. (1996). *Theories of Theories of Mind*. Cambridge: Cambridge University Press.

Cheney, D. L., and R. M. Seyfarth. (1988). Assessment of meaning and the detection of unreliable signals by vervet monkeys. *Animal Behaviour, 36*, 477–86.

———.(1990). *How Monkeys See the World: Inside the Mind of Another Species*. Chicago: Chicago University Press.

Clements, W. A., and J. Perner. (1994). Implicit understanding of belief. *Cognitive Development, 9*, 377–95.

Collins, S. (1993). Is there only one type of male handicap? *Proceedings of the Royal Society, London, 252*, 193–97.

Dunn, J. (1994). Changing minds and changing relationships. In: (Ed. by C. Lewis and P. Mitchell), *Children's early understanding of mind: Origins and development*. (pp. 297–310). Hove: Erlbaum.

Evans, C. S., and P. Marler. (1995). Language and animal communication: Parallels and contrasts. In H. Roitblatt (Ed.), *Comparative Approaches to Cognitive Science*. Cambridge: MIT Press.

Frye, D., and C. Moore. (1991). *Children's Theories of Mind: Mental states and Social Understanding*. Hillsdale: Lawrence Erlbaum.

Gagliardi, J. L., K. K. Kirkpatrick-Steger, J. Thomas, G. J. Allen, and M. S. Blumberg. (1995). Seeing and knowing: Knowledge attribution versus stimulus control in adult humans (*Homo sapiens*). *Journal of Comparative Psychology, 109*, 107–14.

Gergely, G., Z. Nadasdy, G. Csibra, and S. Biro. (1995). Taking the intentional stance at 12 months of age. *Cognition, 56*, 165–93.

Godfray, H. C. J. (1991). Signalling of need by offspring to their parents. *Nature, 352*, 328–30.

Gopnik, A., and A. Meltzoff. (1997). *Words, Thoughts, and Theories*. Cambridge: MIT Press.

Grafen, A. (1990). Biological signals as handicaps. *Journal of Theoretical Biology, 144*, 475–546.

Hala, S., and M. Chandler. (1997). The role of strategic planning in accessing false-belief understanding. *Child Development, 67*, 2948–2966.

Hauser, M. D. (1993). Do vervet monkey infants cry wolf? *Animal Behaviour, 45*, 1242–1244.

———. (1996). *The Evolution of Communication*. Cambridge: MIT Press.

———. (1997). Minding the behavior of deception. In A. Whiten and R. W. Byrne (Eds.), *Machiavellian Intelligence 2* (pp. 112–43). Cambridge: Cambridge University Press.

Hauser, M. D., and P. Marler. (1993a). Food-associated calls in rhesus macaques (*Macaca mulatta*). I. Socioecological factors influencing call production. *Behavioral Ecology, 4*, 194–205.

———. (1993b). Food-associated calls in rhesus macaques (*Macaca mulatta*). II. Costs and benefits of call production and suppression. *Behavioral Ecology, 4*, 206–12.

Heyes, C. M. (1993). Anecdotes, training, trapping and triangulating: Do animals attribute mental states? *Animal Behaviour, 46*, 177–88.

———. (1998). Theory of mind in nonhuman primates. *Behavioral and Brain Sciences, 21*, 101–48.

Johnstone, R. A. (1997). The evolution of animal signals. In J. R. Krebs and N. B. Davies (Eds.), *Behavioural Ecology*. Oxford: Blackwell Scientific.

Johnstone, R. A., and A. Grafen. (1993). Dishonesty and the handicap principle. *Animal Behaviour, 46,* 759–64.

Kilner, R. (1997). Mouth colour is a reliable signal of need in begging canary nestlings. *Proceedings of the Royal Society, London, 264,* 963–68

Leslie, A. M. (1994). ToMM, ToBy, and Agency: Core architecture and domain specificity. In: (Ed. by L. A. Hirschfield and S. A. Gelman), *Mapping the Mind: Domain-specificity in Cognition and Culture.* (pp. 119–48). New York: Cambridge University Press.

Lewis, C., and P. Mitchell. (1994). *Children's Early Understanding of the Mind.* Hillsdale: Lawrence Erlbaum.

Margoliash, D. (1983). Acoustic parameters underlying the responses of song-specific neurons in the white-crowned sparrow. *Journal of Neuroscience, 3,* 1039–1057.

Marler, P., A. Dufty, and R. Pickert. (1986a). Vocal communication in the domestic chicken. I. Does a sender communicate information about the quality of a food referent to a receiver? *Animal Behaviour, 34,* 188–93.

———. (1986b). Vocal communication in the domestic chicken. II. Is a sender sensitive to the presence and nature of a receiver? *Animal Behaviour, 34,* 194–98.

Marler, P., S. Karakashian, and M. Gyger. (1991). Do animals have the option of withholding signals when communication is inappropriate? The audience effect. In C. A. Ristau (Ed.), *Cognitive Ethology: The Minds of Other Animals.* Hillsdale: Lawrence Erlbaum.

Menzel, E. W., and S. Halperin. (1975). Purposive behavior as a basis for objective communication between chimpanzees. *Science, 189,* 652–54.

Møller, A. P. (1993). Morphology and sexual selection in the barn swallow *Hirundo rustica* in Chernobyl, Ukraine. *Proceedings of the Royal Society, London, 252,* 51–57.

———. (1994). *Sexual Selection and the Barn Swallow.* Oxford: Oxford University Press.

Møller, A. P., and J. P. Swaddle. (1997). *Asymmetry, Developmental Stability and Evolution.* Oxford: Oxford University Press.

Munn, C. (1986a). Birds that "cry wolf." *Nature, 319,* 143–45.

———. (1986b). The deceptive use of alarm calls by sentinel species in mixed species flocks of neotropical birds. In R. W. Mitchell and N. S. Thompson (Eds.), *Deception: Perspectives on Human and Nonhuman Deceit.* Albany: State University of New York Press.

Perner, J. (1991). *Understanding the Representational Mind.* Cambridge: MIT Press.

Perrett, D. I., M. H. Harries, A. J. Mistlin, J. K. Hietanen, P. J. Benson, R. Bevan, S. Thomas, M. W. Oram, J. Ortega, and K. Brierly. (1990). Social signals analyzed at the single cell level: Someone is looking at me, something touched me, something moved! *International Journal of Comparative Psychology, 4,* 25–55.

Perrett, D. I., M. W. Oram, M. H. Harries, R. Bevan, J. K. Hietanen, P. J. Benson, and S. Thomas. (1991). Viewer-centered and object-centered coding of heads in the macaque temporal cortex. *Experimental Brain Research, 86,* 159–73.

Povinelli, D. J., K. E. Nelson, and S. T. Boysen. (1990). Inferences about guessing and knowing by chimpanzees (*Pan troglodytes*). *Journal of Comparative Psychology, 104,* 203–10.

———. (1992a). Comprehension of role reversal in chimpanzees: Evidence of empathy? *Animal Behaviour, 43.*

Povinelli, D. J., K. A. Parks, and M. A. Novak. (1991). Do rhesus monkeys (*Macaca mulatta*) attribute knowledge and ignorance to others? *Journal of Comparative Psychology*, 105, 318–25.

———. (1992b). Role reversal by rhesus monkeys, but no evidence of empathy. *Animal Behaviour*, 44, 269–81.

Premack, D., and A. J. Premack. (1995). Origins of human social competence. In M. Gazzaniga (Ed.), *The Cognitive Neurosciences*. (pp. 205–18). Cambridge: MIT Press.

———. (1997). Infants attribute value+/- to the goal-directed actions of self-propelled objects. *Journal of Cognitive Neuroscience*, 9, 848–56.

Reeve, H. K. (1997). Evolutionarily stable communication between kin: A general model. *Proceedings of the Royal Society, London*, 264, 1057–1060.

Ristau, C. (1991a). Aspects of the cognitive ethology of an injury-feigning bird, the piping plover. In C. A. Ristau (Ed.), *Cognitive Ethology: The Minds of Other Animals* Hillsdale: Lawrence Erlbaum.

———. (Ed.). (1991b). *Cognitive Ethology: The Minds of Other Animals*. Hillsdale: Lawrence Erlbaum.

Savage-Rumbaugh, E. S., and R. Lewin. (1996). *Kanzi*. New York: Wiley.

Slaughter, V., and A. Gopnik. (1997). Conceptual coherence in the child's theory of mind: Training children to understand belief. *Child Development*. 67, 2967–2988.

Steger, R., and R. L. Caldwell. (1983). Intraspecific deception by bluffing: A defense strategy of newly molted stomatopods (Arthropods: Crustacea). *Science*, 221, 558–60.

Thornhill, R., and S. Gangestad. (1998). The scent of symmetry: Evidence for a human (*Homo sapiens*) sex pheromone. *Animal Behaviour*, 56, 123–47.

Tovee, M. J., and E. M. Cohen-Tovee. (1993). The neural substrate of face processing models: A review. *Cognitive Neuropsychology*, 10, 505–28.

Van Valen, L. (1962). A study of fluctuating asymmetry. *Evolution*, 16, 125–42.

Viljugrein, H. (1997). The cost of dishonesty. *Proceedings of the Royal Society, London*, 264, 815–21.

Wellman, H. M. (1990). *The Child's Theory of Mind*. Cambridge: Bradford Books, MIT Press.

Wimmer, H., and J. Perner. (1983). Beliefs about beliefs: Representation and constraining function of wrong beliefs in young children's understanding of deception. *Cognition*, 13, 103–28.

Zahavi, A. (1975). Mate selection: a selection for a handicap. *Journal of Theoretical Biology*, 53, 205–14.

Zahavi, A., and A. Zahavi. (1997). *The Handicap Principle*. Oxford: Oxford University Press.

CHAPTER 8

Abler, W. (1989). On the particulate principle of self-diversifying systems. *Journal of Social and Biological Structures*, 12, 1–13.

———. (1997). Gene, language, number: The particulate principle in nature. *Evolutionary Theory*, 11, 237–48.

Beecher, M. D. (1982). Signature systems and kin recognition. *American Zoologist*, 22, 477–90.

Bloom, P. (1994). Possible names: The role of syntax-semantics mappings in the acquisition of nominals. *Lingua*, 92, 297–329.

Bloom, P., and L. Markson. (1998). Capacities underlying word learning. *Trends in Cognitive Science*, 2, 67–73.

Bradbury, J. W., and S. Vehrencamp. (1998). *Animal Communication*. Oxford: Blackwell.

Candland, D. K. (1995). *Feral Children and Clever Animals*. Oxford: Oxford University Press.

Carey, S. (1982). Semantic development: The state of the art. In E. Wanner and L. R. Gleitman (Eds.), *Language Acquisition: The State of the Art* (pp. 347–89). New York: Cambridge University Press.

Carroll, L. (1962). *Alice in Wonderland and Through the Looking Glass*. London: Macmillan.

Cheney, D. L., and R. M. Seyfarth. (1988). Assessment of meaning and the detection of unreliable signals by vervet monkeys. *Animal Behaviour*, 36, 477–86.

———. (1990). *How Monkeys See the World: Inside the Mind of Another Species*. Chicago: University of Chicago Press.

Chomsky, N. (1996). *The Minimalist Program*. Cambridge: MIT Press.

Cleveland, J., and C. T. Snowdon. (1981). The complex vocal repertoire of the adult cotton-top tamarin, *Saguinus oedipus oedipus*. *Zeitschrift für Tierpsychologie*, 58, 231–70.

Colton, R. H., and A. Steinschneider. (1980). Acoustic relationships of infant cries to sudden infant death syndrome. In T. Murry and J. Murry (Eds.), *Infant Communication: Cry and Early Speech* (pp. 183–208). Houston: College-Hill Press.

Crelin, E. (1987). *The Human Vocal Tract*. New York: Vantage Press.

Dawkins, R., and J. R. Krebs. (1978). Animal signals: Information or manipulation. In J. R. Krebs and N. B. Davies (Eds.), *Behavioural Ecology* (pp. 282–309), Oxford: Blackwell Scientific.

Deacon, T. (1997). *The Symbolic Species*. W. W. Norton: New York.

Dennett, D. C. (1983). Intentional systems in cognitive ethology: The "Panglossian paradigm" defended. *Behavioral and Brain Sciences*, 6, 343–90.

———. (1987). *The Intentional Stance*. Cambridge: MIT Press.

Evans, C. S., and P. Marler. (1995). Language and animal communication: Parallels and contrasts. In H. Roitblatt (Ed.), *Comparative Approaches to Cognitive Science* Cambridge: MIT Press.

Evans, R. M. (1994). Cold-induced calling and shivering in young American white pelicans: Honest signalling of offspring need for warmth in a functionally integrated thermoregulatory system. *Behaviour*, 129, 13–34.

Fauconnier, G. (1994). *Mental Spaces*. Cambridge: Cambridge University Press.

Fodor, J. A. (1994). *The Elm and the Expert*. Cambridge: MIT Press.

Fouts, R., and S. T. Mills. (1997). *Next of Kin*. New York: Morrow.

Gardner, R. A., B. T. Gardner, and E. Van Cantfort. (Eds.). (1989). *Teaching Sign Language to Chimpanzees*. Albany: State University of New York Press.

Garstang, M., D. Larom, R. Raspet, and M. Lindeque. (1995). Atmospheric controls on elephant communication. *Journal of Experimental Biology*, 198, 939–51.

Godfray, H. C. J. (1991). Signalling of need by offspring to their parents. *Nature*, 352, 328–30.

Gould, J. L. (1990). Honey bee cognition. *Cognition*, 37, 83–103.

Gould, J. L., and C. G. Gould. (1988). *The Honey Bee*. New York: Freeman Press.

Gouzoules, H., and S. Gouzoules. (1990). Matrilineal signatures in the recruitment screams of pigtail macaques, *Macaca nemestrina. Behaviour, 115,* 327–47.

Green, J. A., and G. E. Gustafson. (1983). Individual recognition of human infants on the basis of cries alone. *Developmental Psychobiology, 16,* 485–93.

Green, S. (1975). Variation of vocal pattern with social situation in the Japanese macaque (*Macaca fuscata*): A field study. In L. A. Rosenblum (Ed.), *Primate Behavior, Volume 4* (pp. 1–102) New York: Academic Press.

Green, S., and P. Marler. (1979). The analysis of animal communication. In P. Marler and J. Vandenbergh (Eds.), *Social Behavior and Communication: Handbook of Behavioral Neurobiology, Volume 3* (pp. 73–158). New York: Plenum.

Griffin, D. R. (1958). *Listening in the Dark.* New Haven: Yale University Press.

Gustafson, G. W., J. A. Green, and T. Tomic. (1984). Acoustic correlates of individuality in the cries of human infants. *Developmental Psychobiology, 17,* 311–24.

Gustafson, G. W., J. A. Green, and J. W. Cleland. (1994). Robustness of individual identity in the cries of human infants. *Developmental Psychobiology, 27,* 1–9.

Hailman, J. P., and M. S. Ficken. (1987). Combinatorial animal communication with computable syntax: Chick-a-dee calling qualifies as "language" by structural linguistics. *Animal Behaviour, 34,* 1899–1901.

Hailman, J. P., M. S. Ficken, and R. W. Ficken. (1985). The "chick-a-dee" calls of *Parus atricapillus:* A recombinant system of animal communication compared with written English. *Semiotica, 56,* 191–224.

———. (1987). Constraints on the structure of combinatorial "chick-a-dee" calls. *Ethology, 75,* 62–80.

Hauser, M. D. (1992). Articulatory and social factors influence the acoustic structure of rhesus monkey vocalizations: A learned mode of production? *Journal of the Acoustic Society of America, 91,* 2175–2179.

———. (1993). Do vervet monkey infants cry wolf? *Animal Behaviour, 45,* 1242–1244.

———. (1996). *The Evolution of Communication.* Cambridge: MIT Press.

———. (1998). Functional referents and acoustic similarity: Field playback experiments with rhesus monkeys. *Animal Behaviour, 55,* 1647–1658.

Hauser, M. D., and R. W. Wrangham. (1987). Manipulation of food calls in captive chimpanzees: A preliminary report. *Folia primatologica, 48,* 24–35.

Herman, L. M., A. A. Pack, and M.-S. Palmer. (1993). Representational and conceptual skills of dolphins. In H. L. Roitblat, L. M. Herman, and P. E. Nachtigall (Eds.), *Language and Communication: Comparative Perspectives* (pp. 403–42). Hillsdale: Lawrence Erlbaum.

Hirsh-Patek, K., and R. M. Golinkoff. (1991). Language comprehension: A new look at some old themes. In N. A. Krasnegor, D. M. Rumbaugh, R. L. Schiefelbusch, and M. Studdert-Kennedy (Eds.), *Biological and Behavioral Determinants of Language Development* (pp. 301–21). Hillsdale: Lawrence Erlbaum.

Huxley, T. H. (1893). *Collected Essays.* London: Macmillan.

Isaac, D., and P. Marler. (1963). Ordering of sequences of singing behaviour of Mistle Thrushes in relationship to timing. *Animal Behaviour, 11,* 179–88.

Jackendoff, R. (1983). *Semantics and Cognition.* Cambridge: MIT Press.

Jackendoff, R. (in press). Possible stages in the evolution of language. *Trends in Cognitive Science.*

Johnstone, R. A. (1997). Recognition and the evolution of distinctive signatures: When does it pay to reveal identity? *Proceedings of the Royal Society, London, 264*, 1547–1553.

Kilner, R. (1997). Mouth colour is a reliable signal of need in begging canary nestlings. *Proceedings of the Royal Society, London, 264*, 963–68.

Lakoff, G. (1987). *Women, Fire, and Dangerous Things*. Chicago: University of Chicago Press.

Langbauer, W. R., Jr., K. Payne, R. Charif, E. Rapaport, and F. Osborn. (1991). African elephants respond to distant playbacks of low frequency conspecific calls. *Journal of Experimental Biology, 157*, 35–46.

Lester, B. M., and C. F. Z. Boukydis. (1992). No language but a cry. In H. Papousek, U. Jürgens, and M. Papousek (Eds.), *Nonverbal Vocal Communication: Comparative and Developmental Approaches* (pp. 145–73). Cambridge: Cambridge University Press.

Lieberman, P. (1984). *The Biology and Evolution of Language*. Cambridge: Harvard University Press.

———. (1985). The physiology of cry and speech in relation to linguistic behavior. In B. M. Lester and C. F. Z. Boukydis (Eds.), *Infant Crying* (pp. 29–57). New York: Plenum.

Macedonia, J. (1991). What is communicated in the antipredator calls of lemurs: Evidence from playback experiments with ring-tailed and ruffed lemurs. *Ethology, 86*, 177–90.

Macnamara, J. (1982). *Names for Things*. Cambridge: MIT Press.

Markman, E. M. (1990). Constraints children place on word meanings. *Cognitive Science, 14*, 57–77.

Markman, E. M., and J. E. Hutchinson. (1984). Children's sensitivity to constraints on word meaning: Taxonomic versus thematic relations. *Cognitive Psychology, 16*, 1–27.

Marler, P. (1976). Social organization, communication and graded signals: The chimpanzee and the gorilla. In P. P. G. Bateson and R. A. Hinde (Eds.), *Growing Points in Ethology* (pp. 239–80). Cambridge: Cambridge University Press.

———. (1977). The structure of animal communication sounds. In T. H. Bullock (Ed.), *Recognition of Complex Acoustic Signals* (pp. 17–35). Berlin: Springer-Verlag.

Marler, P., A. Dufty, and R. Pickert. (1986a). Vocal communication in the domestic chicken. I. Does a sender communicate information about the quality of a food referent to a receiver? *Animal Behaviour, 34*, 188–93.

———. (1986b). Vocal communication in the domestic chicken. II. Is a sender sensitive to the presence and nature of a receiver? *Animal Behaviour, 34*, 194–98.

Marler, P., and S. Peters. (1989). Species differences in auditory responsiveness in early vocal learning. In R. J. Dooling and S. H. Hulse (Eds.), *The Comparative Psychology of Audition: Perceiving Complex Sounds* (pp. 243–73). Hillsdale: Lawrence Erlbaum.

Marler, P., S. Peters, G. F. Ball, A. M. Dufty, and J. C. Wingfield. (1988). The role of sex steroids in the acquisition and production of birdsong. *Nature, 336*, 770–72.

Marler, P., and R. Pickert. (1984). Species-universal microstructure in the learned song of the swamp sparrow (*Melospiza georgiana*). *Animal Behaviour, 32*, 673–89.

Matsuzawa, T. (1996). Chimpanzee intelligence in nature and in captivity: Isomorphism of symbol use and tool use. In W. C. McGrew, L. F. Nishida, and T. Nishida (Eds.), *Great Ape Societies* (pp. 196–209). Cambridge: Cambridge University Press.

McGowan, B., and D. Reiss. (1997). Vocal learning in bottlenose dolphins: A comparison of humans and nonhuman animals. In C. T. Snowdon and M. Hausberger (Eds.), *Social Influences on Vocal Development* (pp. 178–207). Cambridge: Cambridge University Press.

Michelsen, A. (1992). Hearing and sound communication in small animals: Evolutionary adaptations to the laws of physics. In D. B. Webster, R. F. Fay, and A. N. Popper (Eds.), *The Evolutionary Biology of Hearing* (pp. 61–77). New York: Springer-Verlag.

Michelsen, A., B. B. Andersen, J. Storm, W. H. Kirchner, and M. Lindauer. (1992). How honeybees perceive communication dances, studied by means of a mechanical model. *Behavioral Ecology and Sociobiology, 30,* 143–50.

Miller, G. (1996). *The Science of Words.* New York: New York.

Negus, V. E. (1949). *The Comparative Anatomy and Physiology of the Larynx.* New York: Hafner.

Patterson, F. (1987). *Koko's Story.* New York: Scholastic.

Payne, K. B., and R. S. Payne. (1985). Large scale changes over 19 years in songs of humpback whales in Bermuda. *Zeitschrift für Tierpsychologie, 68,* 89–114.

Payne, K.,Tyack, P., and R. Payne. (1984). Progressive changes in the songs of humpback whales (*Megaptera novaeangliae*): A detailed analysis of two seasons in Hawaii. In R. Payne (Ed.), *Communication and Behavior of Whales* (pp. 9–57), Boulder: Westview.

Pepperberg, I. M. (1991). A communicative approach to animal cognition: A study of conceptual abilities of an African grey parrot. In C. A. Ristau (Ed.), *Cognitive Ethology: The Minds of Other Animals* (pp. 153–86). Hillsdale: Lawrence Erlbaum.

Pinker, S. (1994). *The Language Instinct.* New York: William Morrow.

Premack, D. (1986). *Gavagai! or the Future History of the Animal Language Controversy.* Cambridge: MIT Press.

———. (1990). Words: What are they, and do animals have them? *Cognition, 37,* 197–212.

Premack, D., and A. Premack. (1983). *The Mind of an Ape.* New York: W. W. Norton.

Quine, W. V. 0. (1960). *Word and Object.* Cambridge: MIT Press.

———. (1973). On the reasons for the indeterminacy of translation. *Journal of Philosophy, 12,* 178–83.

Redington, M., and N. Chater. (1997). Probabilistic and distributional approaches to language acquisition. *Trends in Cognitive Science, 1,* 273–81.

Rendall, D., P. S. Rodman, and R. E. Edmond. (1996). Vocal recognition of individuals and kin in free-ranging rhesus monkeys. *Animal Behaviour, 51,* 1007–1015.

Robinson, J. G. (1979). An analysis of the organization of vocal communication in the titi monkey *Callicebus moloch. Zeitschrift für Tierpsychologie, 49,* 381–405.

———. (1984). Syntactic structures in the vocalizations of wedge-capped capuchin monkeys, *Cebus nigrivittatus. Behaviour, 90,* 46–79.

Saffran, J. R., R. N. Aslin, and E. L. Newport. (1996). Statistical cues in language acquisition: Word segmentation by infants. In G. W. Cottrell (Ed.), *Proceedines of the 18th Annual Conference of the Cognitive Science Society* (pp 376–80). Hillsdale: Lawrence Erlbaum.

Savage-Rumbaugh, E. S. (1986). *Ape Language: From Conditioned Response to Symbol.* New York: Columbia University Press.

Savage-Rumbaugh, E. S., and R. Lewin. (1996). *Kanzi.* New York: Wiley.

Schusterman, R. J., R. Gisiner, B. K. Grimm, and E. B. Hanggi. (1993). Behavior control by exclusion and attempts at establishing semanticity in marine mammals using match-to-sample paradigms. In H. L. Roitblat, L. M. Herman, and P. E. Nachtigall (Eds.), *Language and Communication: Comparative Perspectives* (pp. 249–74). Hillsdale: Lawrence Erlbaum.

Sebeok, T. A. (1977). *How Animals Communicate*. Bloomington: Indiana University Press.

Simmons, J. (1990). A view of the world through the bat's ear: The formation of acoustic images in echolocation. *Cognition, 33,* 155–99.

Snowdon, C. T., and M. Hausberger. (Eds.) (1997). *Social Influences on Vocal Development*. Cambridge: Cambridge University Press.

Struhsaker, T. T. (1967). Auditory communication among vervet monkeys (*Cercopithecus aethiops*). In S. A. Altmann (Ed.), *Social Communication among Primates* (pp. 281–324). Chicago: University of Chicago Press.

Studdert-Kennedy, M. (1998). The particulate origins of language generativity: From syllable to gesture. In J. Hurford, M. Studdert-Kennedy, and C. Knight (Eds.), *Approaches to the Evolution of Language: Social and Cognitive Bases* (pp. 202–21). Cambridge: Cambridge University Press.

Tomasello, M. (1988). The role of joint attentional processes in early language development. *Language Sciences, 10,* 69–88.

Trivers, R. L. (1972). Parental investment and sexual selection. In B. Campbell (Ed.), *Sexual Selection and the Descent of Man* (pp. 136–79). Chicago: Aldine.

———. (1974). Parent-offspring conflict. *American Zoologist, 14,* 249–64.

Tyack, P. L. (1993). Animal language research needs a broader comparative evolutionary framework. In H. L. Roitbtat, L. M. ,Herman, and P. E. Nachtigall (Eds.), *Language and Communication: Comparative Perspectives* (pp. 115–52). Hillsdale: Lawrence Erlbaum.

von Frisch, K. (1967). *The Dance Language and Orientation of Bees*. Cambridge: Belknap Press of Harvard University Press.

Wallman, J. (1992). *Aping Language*. New York: Cambridge University Press.

Wittgenstein, L. (1958). *Philosophical Investigations*. Oxford: Blackwell.

Zeskind, P. S., J. Sale, M. L. Maio, L. Huntington, and J. R. Weiseman. (1985). Adult perceptions of pain and hunger cries: A synchrony of arousal. *Child Development, 56,* 549–54.

Zuberbuhler, K., R. Noe, and R. M. Seyfarth. (1997). Diana monkey long-distance calls: Messages for conspecifics and predators. *Animal Behaviour, 53,* 589–604.

CHAPTER 9

Allen, C., and M. D. Hauser. (1991). Concept attribution in nonhuman animals: Theoretical and methodological problems in ascribing complex mental processes. *Philosophy of Science, 58,* 221–40.

Anders, T. (1994). *The Evolution of Evil*. Chicago: Open Court Publishing.

Axelrod, R. (1984). *The Evolution of Cooperation*. New York: Basic Books.

Bachmann, C., and H. Kummer. (1980). Male assessment of female choice in hamadryas baboons. *Behavioral Ecology and Sociobiology, 6,* 315–21.

Baillargeon, R. (1995). Physical reasoning in infancy. In M. Gazzaniga (Ed.), *The Cognitive Neurosciences*. (pp. 181–204). Cambridge: MIT Press.

Baldwin, D. A., and L. M. Moses. (1996). The ontogeny of social information gathering. *Child Development*, 67, 1915–1939.

Barkow, J., L. Cosmides, and J. Tooby. (Eds.). (1992). *The Adapted Mind*. New York: Oxford University Press.

Barton, R. A., and R. I. M. Dunbar. (1997). Evolution of the social brain. In A. Whiten and R. W. Byrne (Eds.), *Machiavellian Intelligence 2* (pp. 240–63). Cambridge: Cambridge University Press.

Beach, F. A., M. G. Buehler, and I. F. Dunbar. (1982). Competitive behavior in male, female, and pseudohermaphroditic female dogs. *Journal of Comparative Physiological Psychology*, 96, 855–74.

Binmore, K. (1992). *Fun and Games: A Text on Game Theory*. Lexington: Heath.

———. (1994). *Game Theory and the Social Contract: Playing Fair*. Cambridge: MIT Press.

———. (1998). *Game Theory and the Social Contract: Just Playing*. Cambridge: MIT Press.

Boyd, R. (1992). The evolution of reciprocity when conditions vary. In A. H. Harcourt and F. B. M. de Waal (Eds.), *Coalitions and Alliances in Humans and Other Animals* Oxford: Oxford University Press.

Campos, J. J., and C. R. Stenberg. (1981). Perception, appraisal and emotion: The onset of social referencing. In M. E. Lamb and L. R. Sherrod (Eds.), *Infant Social Cognition*, Hillsdale: Lawrence Erlbaum.

Carey, S. (1985). *Conceptual Change in Childhood*. Cambridge: MIT Press.

Cavalieri, P., and P. Singer. (1993). *The Great Ape Project*. New York: St. Martin's Press.

Cheney, D. L., and R. M. Seyfarth. (1980). Vocal recognition in free-ranging vervet monkeys. *Animal Behaviour*, 28, 362–67.

Chevalier-Skolnikoff , S. (1973). Facial expression of emotion in nonhuman primates. In P. Ekman (Ed.), *Darwin and Facial Expression* (pp. 11–90). London: Academic Press.

Church, R. M. (1959). Emotional reactions of rats to the pain of others. *Journal of Comparative and Physiological Psychology* 52, 132–34.

Clark, S. R. L, (1997). *Animals and Their Moral Standing*. London: Routledge.

Clutton-Brock, T. H., and G. A. Parker. (1995). Punishment in animal societies. *Nature*, 373, 209–16.

Connor, R. C., R. A. Smolker, and A. F. Richards. (1992). Dolphin alliances and coalitions. In A. H. Harcourt and F. B. M. de Waal (Eds.), *Coalitions and Alliances in Humans and Other Animals* (pp. 415–43). Oxford: Oxford University Press.

Cords, M., and F. Aureli. (1993). Patterns of reconciliation among juvenile long-tailed macaques. In M. E. Pereira and L. A. Fairbanks (Eds.), *Juvenile Primates: Life History, Development, and Behavior* (pp. 271–84). New York: Oxford University Press.

Cords, M., and S. Thurnheer. (1993). Reconciling with valuable partners by long-tailed macaques. *Ethology*, 93, 315–25.

Cosmides, L., and J. Tooby. (1992). Cognitive adaptations for social exchange. In J. Barkow, L. Cosmides, and J. Tooby (Eds.), *The Adapted Mind* (pp. 163–228). New York: Oxford University Press.

———. (1994). Origins of domain specificity: The evolution of functional organization. In L. A. Hirschfeld and S. A. Gelman (Eds.), *Mapping the Mind: Domain-Specificity in Cognition and Culture* (pp. 85–116). Cambridge: Cambridge University Press.

Daly, M., and M. Wilson. (1988). *Homicide*. New York: Aldine de Gruyter.

Damasio, A. (1994). *Descartes' Error: Emotion, Reason and the Human Brain*. New York: Putnam.

Darwin, C. (1871). *The Descent of Man and Selection in Relation to Sex*. London: John Murray.

Davidson, R. J. (1995). Cerebral asymmetry, emotion, and affective style. In R. J. Davidson and K. Hugdahl (Eds.), *Brain Asymmetry* (pp. 361–89). Cambridge: MIT Press.

Dawkins, M. S. (1983). Battery hens name their price: Consumer demand theory and the measurement of ethological "needs." *Animal Behaviour, 31*, 1195–1205.

———. (1990). From an animal's point of view: Motivation, fitness and animal welfare. *Behavioral and Brain Sciences, 13*, 1–61.

Dawkins, R. (1976). *The Selfish Gene*. Oxford: Oxford University Press.

———. (1986). *The Blind Watchmaker*. New York: Norton.

Deacon, T. W. (1997). *The Symbolic Species: The Co-evolution of Language and the Brain*. New York: Norton.

Dennett, D. (1987). *The Intentional Stance*. Cambridge: MIT Press.

———. (1995). *Darwin's Dangerous Idea: Evolution and the Meanings of Life*. New York: Simon and Schuster.

de Waal, F. B. M. (1982). *Chimpanzee Politics: Power and Sex Among Apes*. New York: Harper and Row.

———. (1989a). Food sharing and reciprocal obligations among chimpanzees. *Journal of Human Evolution, 18*, 433–59.

———. (1989b). *Peacemaking among Primates*. Cambridge: Cambridge University Press.

———. (1996). *Good Natured*. Cambridge: Harvard University Press.

de Waal, F. B. M., and D. L. Johanowicz. (1993). Modification of reconciliation behavior through social experience: An experiment with two macaque species. *Child Development, 64*, 897–908.

de Waal, F. B. M., and A. van Roosmalen. (1979). Reconciliation and consolation among chimpanzees. *Behavioral Ecology and Sociobiology, 5*, 55–66.

de Waal, F. B. M., and D. Yoshihara. (1983). Reconciliation and redirected aggression in rhesus monkeys. *Behaviour, 85*, 224–41.

Diamond, A. (1988). Differences between adult and infant cognition: Is the crucial variable presence or absence of language? In L. Weiskrantz (Ed.), *Thought without Language* (pp 337–70). Oxford: Clarendon Press.

Dissanayake, E. (1992). *Homo Aestheticus: Where Art Comes From and Why*. Seattle: University of Washington Press.

Dugatkin, L. A. (1997). *Cooperation among Animals: An Evolutionary Perspective*. New York: Oxford University Press.

Dunbar, R. (1996). *Grooming, Gossip and the Evolution of Language*. Cambridge: Harvard University Press.

Ekman, P., R. W. Levenson, and W. V. Friesen. (1983). Autonomic nervous system activity distinguishes among emotions. *Science, 218*, 1208–1210.

Fischer, J. M. (1993). *The Metaphysics of Death*. Stanford: Stanford University Press.

Fridlund, A. (1994). *Human Facial Expressions: An Evolutionary Perspective*. New York: Academic Press.

Gibbard, A. (1990). *Wise Choices, Apt Feelings*. Cambridge: Harvard University Press.

Goldman-Rakic, P. S. (1987). Circuitry of primate prefrontal cortex and regulation of behavior by representational memory. In F. Plum (Ed.), *Handbook of Physiology, Section 1, The Nervous System, Volume 5, Higher Functions of the Brain* (pp. 373–417). Bethesda: American Physiological Society.

Gopnik, A., and A. Meltzoff. (1997). *Words, Thoughts, and Theories*. Cambridge: MIT Press.

Griffiths, P. E. (1997). *What Emotions Really Are*. Chicago: University of Chicago Press.

Hamilton, W. D. (1964a). The evolution of altruistic behavior. *American Naturalist, 97*, 354–56.

———. (1964b). The genetical evolution of social behavior. *Journal of Theoretical Biology, 7*, 1–52.

———. (1971). Geometry for the selfish herd. *Journal of Theoretical Biology, 31*, 295–311.

Harcourt, A. H. (1988). Alliances in contests and social intelligence. In R. Byrne and A. Whiten (Eds.), *Machiavellian intelligence*, (pp. 132–52). Oxford: Clarendon Press.

Harcourt, A. H., and F. B. M. de Waal (Eds.). (1992). *Coalitions and Alliances in Humans and Other Animals*. Oxford: Oxford University Press.

Hatano, G., and K. Inagaki. (1995). Young children's naïve theory of biology. *Cognition, 50*, 153–70.

Hauser, M. D. (1996). *The Evolution of Communication*. Cambridge: MIT Press.

———. (1997). Minding the behavior of deception. In A. Whiten and R. Byrne (Eds.), *Machiavellian Intelligence 2* (pp. 112–43). Cambridge: Cambridge University Press.

———. (1998). Expectations about object motion and destination: Experiments with a nonhuman primate. *Developmental Science, 1*, 31–38.

Heider, F., and M. Simmel. (1944). An experimental study of apparent behavior. *American Journal of Psychology, 57*, 243–59.

Heinsohn, R., and C. Packer. (1995). Complex cooperative strategies in group-territorial African lions. *Science, 269*, 1260–1262.

Hemelrijk, C. K. (1990). Models of, and tests for, reciprocity, unidirectionality and other social interaction patterns at a group level. *Animal Behaviour, 39*, 1013–1029.

Hood, B. (1995). Gravity rules for 2-4-year-olds? *Cognitive Development, 10*, 577–98.

Hood, B. M., M. D. Hauser, L. Anderson, and L. Santos. (1999). Gravity biases in a non-human primate? *Developmental Science, 2*, 35–41.

Huber, E. (1931). *Evolution of the Facial Musculature and Facial Expression*. Baltimore: Johns Hopkins University Press.

Keil, F. C. (1994). The birth and nurturance of concepts by domains: The origins of concepts of living things. In L. A. Hirschfeld and S. A. Gelman (Eds.), *Mapping the Mind: Domain-Specificity in Cognition and Culture* (pp. 234–54). Cambridge: Cambridge University Press.

Kuhn, T. S. (1970). *The Structure of Scientific Revolutions*. Chicago: University of Chicago Press.

Kummer, H. (1978). On the value of social relationships to nonhuman primates: A heuristic scheme. *Social Science Information, 17*, 687–705.

Kummer, H., and M. Cords. (1991). Cues of ownership in *Macaca fasicularis*. *Animal Behaviour, 42*, 529–49.

Kummer, H., W. Gotz, and W. Angst. (1974). *Triadic differentiation: An inhibitory process protecting pair bonds in baboons. Behaviour, 49*, 62–87.

Lakatos, I. (1978). *Philosophical Papers; Volume 1, The Methodology of Scientific Research Programmes*. Cambridge: Cambridge University Press.

Lavery, J. J., and P. J. Foley. (1963). Altruism or arousal in the rat? *Science, 140,* 172–73.

LeDoux, J. (1996). *The Emotional Brain: The Mysterious Underpinnings of Emotional Life*. New York: Simon and Schuster.

Leslie, A. M. (1982). The perception of causality in infants. *Perception, 11,* 173–86.

———. (1994). ToMM, ToBy, and Agency: Core architecture and domain specificity. In L. A. Hirschfeld and S. A. Gelman (Eds.), *Mapping the Mind: Domain-specificity in Cognition and Culture* (pp. 119–48). Cambridge: Cambridge University Press.

Leslie, A. M., and S. Keeble. (1987). Do six-month old infants perceive causality? *Cognition, 25,* 265–88.

Mason, G., J. Garner, and D. McFarland. (1998). Assessing animal priorities: Future directions. *Animal Behaviour, 55,* 1082–1083.

Masserman, J. H., S. Wechkin, and T. William. (1964). Altruistic behavior in rhesus monkeys. *American Journal of Psychiatry, 121,* 584–85.

Masson, J. M., and S. McCarthy. (1995). *When Elephants Weep*. New York: Delacorte.

May, L., M. Friedman, and A. Clark. (1996). *Minds and Morals*. Cambridge: MIT Press.

Maynard Smith, J. (1964). Group selection and kin selection. *Nature, 201,* 1145–1147.

———. (1965). The evolution of alarm calls. *American Naturalist, 99,* 59–63.

———. (1974). The theory of games and the evolution of animal conflicts. *Journal of Theoretical Biology, 47,* 209–21.

Michotte, A. (1962). *The Perception of Causality*. Andover: Methuen.

Miller, R. E. (1967). Experimental approaches to the physiological and behavioral concomitants of affective communication in rhesus monkeys. In S. A. Altmann (Ed.), *Social Communication among Primates*. Chicago: University of Chicago Press.

Miller, R. E., J. Banks, and H. Kuwahara. (1966). The communication of affect in monkeys: Cooperative conditioning. *Journal of Genetic Psychology, 108,* 121–34.

Miller, R. E., J. Banks, and N. Ogawa. (1962). Communication of affect in "cooperative conditioning" of rhesus monkeys. *Journal of Abnormal Social Psychology, 64,* 343–48.

———. (1963). The role of facial expression in "cooperative-avoidance" conditioning in monkeys. *Journal of Abnormal Social Psychology, 67,* 24–30.

Moss, C. (1988). *Elephant Memories*. New York: William Morrow.

Pagels, E. (1988). *Adam, Eve, and the Serpent*. New York: Vintage Books.

Petrinovich, L. (1995). *Human Evolution, Reproduction, and Morality*. New York: Plenum.

———. (1999). *Darwinian Dominion: Animal Welfare and Human Interests*. Cambridge: MIT Press.

Pinker, S. (1997). *How the Mind Works*. New York: Norton.

Posner, R. (1992). *Sex and Reason*. Cambridge: Harvard University Press.

Premack, D., and A. J. Premack. (1995). Origins of human social competence. In M. Gazzaniga (Ed.), *The Cognitive Neurosciences* (pp. 205–18). Cambridge: MIT Press.

Ridley, M. (1996). *The Origins of Virtue*. New York: Viking Press/Penguin Books.

Rollin, B. E. (1989). *The Unheeded Cry*. Oxford: Oxford University Press.

Rottschaefer, W. A. (1998). *The Biology and Psychology of Moral Agency*. Cambridge: Cambridge University Press.

Russel, J., N. Mauthner, S. Sharpe, and T. Tidswell. (1991). The "windows task" as a measure of strategic deception in preschoolers and autistic subjects. *British Journal of Developmental Psychology*, 9, 331–49.

Santos, L. R., B. Ericson, and M. D. Hauser. (1999). Constraints on problem solving and inhibition: Object retrieval in cotton-top tamarins. *Journal of Comparative Psychology*, 113, 186–93.

Sigg, H., and J. Falett. (1985). Experiments on the respect of possession and property in hamadryas baboons (*Papio hamadryas*). *Animal Behaviour*, 33, 978–84.

Silberberg, A., and K. Fujita. (1996). Pointing at smaller food amounts in an analogue of Boysen and Bernston's (1995) procedure. *Quarterly Journal of Experimental Psychology*, 66, 143–47.

Silk, J. B. (in press). Making amends: Adaptive perspectives on conflict remediation in monkeys, apes, and humans. *Human Nature*.

Sober, E., and D. S. Wilson. (1998). *Unto Others*. Cambridge: Harvard University Press.

Spelke, E. S., A. T. Phillips, and A. L. Woodward. (1995). Infants' knowledge of object motion and human action. In A. J. Premack, D. Premack, and D. Sperber (Eds.), *Causal Cognition: A Multidisciplinary Debate* (pp. 44–77). Oxford: Clarendon Press.

Stammbach, E. (1988). Group responses to specially skilled individuals in a *Macaca fascicularis. Behaviour*, 107, 241–66.

Trivers, R. L. (1971). The evolution of reciprocal altruism.*Quarterly Review of Biology*, 46, 35–57.

———. (1972). Parental investment and sexual selection. In B. Campbell (Ed.), *Sexual Selection and the Descent of Man* (pp. 136–79). Chicago: Aldine.

———. (1974). Parent-offspring conflict. *American Zoologist*, 14, 249–64.

Wechkin, S., J. H. Masserman, and W. Terris, Jr. (1964). Shock to a conspecific as an aversive stimulus. *Psychonomic Science*, 1, 47–48.

Wilkinson, G. S. (1984). Reciprocal food sharing in the vampire bat. *Nature*, 308, 181–84.

Williams, G. C. (1966). *Adaptation and Natural Selection*. Princeton: Princeton University Press.

Woodward, A. L. (1998). Infants selectively encode the goal object of an actor's reach. *Cognition*, 69, 1–31.

Wright, R. (1994). *The Moral Animal*. New York: Pantheon.

INDEX